P9-DDK-715

ACCIDENTAL
EXPLORERS

EXTRAORDINARY EXPLORERS

ACCIDENTAL EXPLORERS
Surprises and Side Trips in the History of Discovery

WOMEN OF THE WORLD
Women Travelers and Explorers

SCIENTIFIC EXPLORERS
Travels in Search of Knowledge

ACCIDENTAL EXPLORERS

Surprises and Side Trips in the History of Discovery

Rebecca Stefoff

Oxford University Press
New York • Oxford

To Don C. and Suzanne Stefoff,
with love and gratitude

Oxford University Press

Oxford New York
Athens Auckland Bangkok Bogotá Bombay
Buenos Aires Calcutta Cape Town Dar es Salaam
Delhi Florence Hong Kong Istanbul Karachi
Kuala Lumpur Madras Madrid Melbourne
Mexico City Nairobi Paris Singapore
Taipei Tokyo Toronto
and associated companies in
Berlin Ibadan

Published by Oxford University Press, Inc.,
198 Madison Avenue, New York, New York 10016

Design: Charles Kreloff
Picture research: Wendy P. Wills

Library of Congress Cataloging-in-Publication Data

Stefoff, Rebecca
Accidental explorers : surprises and sidetrips in the history of exploration / Rebecca Stefoff.
p. cm. — (Extraordinary explorers)
Includes bibliographical references.
Summary: Discusses the role of chance in the discoveries of such explorers as Columbus,
Ponce de León, Dr. David Livingstone, and Jedediah Smith.
ISBN 0-19-507685-0
1. Discoveries in geography—juvenile literature.
[1. Discoveries in geography. 2. Explorers.]
I. Title. II. Series: Stefoff, Rebecca, Extraordinary explorers.
G175.S83 1992
910'.9—dc20 92-5211
CIP
AC

3 5 7 9 8 6 4 2

Printed in the United States of America
on acid-free paper

On the cover: Geradus Mercator, from the frontispiece of his *Atlas sive Cosmographicae* (1585-95)

Contents

INTRODUCTION

Walpole's Wonderful Word

There is an island in the Indian Ocean that hangs like a pear-shaped pendant from the tip of the great peninsula of India. This island has long been the source of many of the world's most prized gems. Blood-red rubies, glittering dark-blue sapphires, and tawny yellow topazes are mined from its mountains, and the warm tropical waters around it yield lustrous pearls. For more than a thousand years, the island lured treasure hunters, traders, and explorers across the perilous southern seas.

Since 1972 the island has been called Sri Lanka. For centuries before that, it was called Ceylon. Earlier still, the Greek- and Latin-speaking peoples of the ancient world called it Taprobane, a name that is now forgotten except by scholars. But the Arabs who sailed the Indian Ocean in medieval times had another name for it. They called the island Serendip, and that name has lived on into modern times in a surprising way, all because of a Persian fairy tale called *The Three Princes of Serendip.*

Horace Walpole, an eighteenth-century British writer, read this tale, which tells the story of three accident-prone princes who never achieved what they set out to do, but whose wrong turns and blunders always turned out to benefit them in the end. As Walpole wrote in a 1754 letter, the princes of Serendip "were always making discoveries, by accidents and sagacity, of things they were not in quest of."

Walpole invented the word *serendipity* to describe the making of a fortunate or unexpected discovery by accident, and the word made its way into the English language. A writer who used the word in 1880 defined it as "looking for one thing and finding another"—an excellent description of many milestones in human history.

Enterprises such as scientific research, voyages of discovery, or even the writing of books do not always proceed smoothly and logically from A to B to C. Sometimes the scientist or traveler or writer finds that there *is* no C. Sometimes he gets off course and unexpectedly emerges at J. And sometimes he reaches C after all, only to find that it is not quite what he had imagined it to be. The history of discovery is filled with stories of serendipitous finds, wrong turns, accidents, and surprises. Many travelers and adventurers have added to our knowledge of the world in unplanned ways.

Some of these journeyers have been enshrined in history as heroes of exploration; others have been almost forgotten. Some of them made enormous, world-changing discoveries; others had obscure adventures that are nothing more than colorful footnotes to history. Some of them lived in ancient times; others are modern. But each of them, in one way or another, enlarged or enriched our view of the world—often without meaning to. They are the real princes and princesses of Serendip, the accidental explorers.

PART ONE

DISCOVERED BY ACCIDENT

Christopher Columbus, who bumped repeatedly into the inconveniently located Americas while trying to sail west from Europe to Asia, is the best-known accidental explorer of all. But there were others, before him and after, who set out to look for one thing and found something else entirely.

CHAPTER 1

Viking Luck

These ninth-century silver coins show a Viking knörr (top), used for trading, and a longship (bottom), used in war.

Five hundred years before Columbus, there were Europeans in the New World. They explored, they built homes, and they dealt in both friendship and war with the native peoples of North America. These Europeans were the Vikings, and they stumbled onto America by what historian David Divine, in his book *The Opening of the World,* calls "a succession of stupendous accidents." The accidents involved a series of Viking mariners who had the bad luck to get lost—and the good luck to find their way home again.

The Vikings, or Norsemen, were Scandinavian seafarers from Norway, Denmark, and Sweden. In the eighth century A.D., they burst out of their homelands on a centuries-long spree of raiding, trading, and colonizing. To the east, they sailed down the rivers of Russia to the Black Sea and the Caspian Sea. To the south, they conquered parts of England, Ireland, France, Spain, and Italy. In all of these places, they sailed close to land. But in the west, they ventured out of sight of land, deep into the ocean realm, until eventually they were spread across the chain of islands that stretches across the North Atlantic.

The Vikings were master shipbuilders who developed two main types of ships. The longship, sometimes called the dragon ship, was long and lean and swift; it was a warship that could run onto a beach or up a river mouth in a lightning raid on an unsuspecting village. The *knörr* was shorter, but it had higher sides and rode deeper in the water. It was better suited than the longship for long deep-water voyages in rough seas, so it was the knörr that carried the Vikings across the stormy Atlantic Ocean.

The Vikings' first serious assault on Europe was an attack on a monastery on Lindisfarne Island, off the English coast, in 793.

Carved in wood by a Norse craftsman in the early ninth century, a Viking glares at the world (opposite).

Throughout the following century, while they were pillaging and plundering the European coasts in their dreaded longships, the Vikings were also sailing in their sturdy knörrs to the island clusters that dot the Atlantic Ocean north of the British Isles: the Shetlands, the Orkneys, and the Faeroes. Norse colonies were established on these outposts. The most remote were the Faeroe Islands, destined to be the springboard for the accidental discovery of North America.

The first of the Vikings' three big accidents occurred sometime in the middle of the ninth century. An outlaw known only as Naddod was on his way from Norway to the Faeroe Islands when his knörr was caught in a gale and driven north and west of the Faeroes. Days later, when the storm cleared, Naddod saw land ahead of him in the west. He landed in a region of snowy peaks and climbed one of them before making his way back to the Faeroes. He gave the name Snowland to the place he had found.

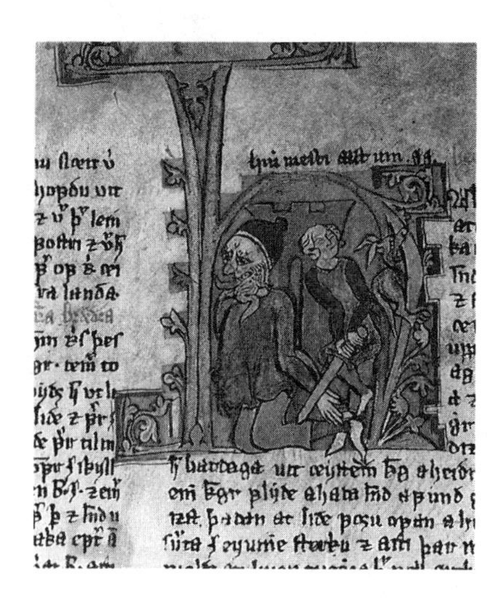

Harald Fairhair, king of Norway, frees a giant from bondage in a scene from Norse legend. Many Norwegians moved to Iceland to escape Harald's harsh rule.

At about the same time, a Swede named Gardar Svavarsson was also driven off course by bad weather, and he happened upon the same land. He sailed all the way around it, thus discovering that it was an island, and he noted the smoke that rose from its rumbling peaks and the long rivers of ice that wound down from its mountains to the sea. Upon returning home, Svavarsson told of his discovery, calling the island Gardarsholm.

A Viking named Floki Vilgerdason then went looking for this rumored island of volcanoes and glaciers in the cold gray northern sea. He found it and gave it the name it still bears today: Iceland. The first permanent Norse colony was established in Iceland in 874. Vikings from Norway eagerly moved to the new land, partly because their homeland was growing crowded, but mostly to escape the tyrannical rule of Harald Fairhair and other stern kings who ruled Norway in the late ninth and early tenth centuries. By the middle of the tenth century, Iceland had been extensively settled, and its habitable areas were dotted with the stone-and-sod homesteads of the Norsemen. But by that time the Vikings had made their second great accidental discovery, and once again it happened when a mariner was blown off course.

Some of the Norse sagas, or folk histories, give the year of discovery as 900; others say it was 930. Gunnbjorn Ulfsson was on his way from Norway to Iceland with a cargo of goods for the Icelanders when his knörr was seized in the grip of powerful winds from the northeast. The gale carried him some days' sail west of Iceland, until, in the sea ahead of him, he spied rocks breaking through the wild waves. A wall of ice loomed up in the distance. Fearful of crashing upon what seemed a distinctly unwelcoming shore, Ulfsson managed to turn around and make his way back to Iceland, where he told everyone what he had seen. He called the place Gunnbjorn's Skerry—a skerry is a rocky islet or reef—and his voyage passed into Icelandic folklore.

At the westernmost edge of the known world, Iceland became a refuge for people who could not remain in their native countries. One of these was a man named Thorvald Asvaldsson, who was exiled from Norway for killing a man. He arrived in Iceland in about 960, accompanied by his son Erik Thorvaldsson, who was called Erik the Red because he had red hair. Erik and his family took the lead roles in the drama of the Vikings in America.

Like his father, Erik was a man of quarrelsome and violent temper. He killed a man in a dispute, and the Icelandic parliament ordered him to move to an isolated settlement on one of the island's western peninsulas. He slew another man a few years later, and this time he was ordered to leave Iceland altogether for three years. He decided to spend this period of banishment in a voyage west, in search of Gunnbjorn's Skerry, that unknown land that Ulfsson had glimpsed some fifty or eighty years before. Perhaps it would be another Iceland, just waiting for him to claim it.

Erik set off in the spring of 982 with a knörr full of supplies and a crew made up of his clan's followers. About 500 miles (800 kilometers) west of Iceland they came upon land: first the rocky reefs that Ulfsson had seen, and beyond them a grim, high coast of ice and rock. Heading south, Erik skirted this coast until he passed around a cape and turned north. Now he was sailing along the western coast of the new land, which appeared much more welcom-

Early sailors believed the ocean was full of mysteries and perils. Olaus Magnus, a sixteenth-century Swedish scholar, illustrated his map of the northern seas with this mythical whale that was said to devour ships and their crews.

ing than the eastern. This western coast, in fact, seemed surprisingly familiar to the Norsemen. It was cut by many narrow, winding, rock-walled inlets of the sea, very much like the fjords of Norway. Like Norway it had a zone of grassy meadows between the sea and the snow-covered highlands. The waters teemed with fish and seals, the caribou that roamed the land were similar to the reindeer of Norway, and the millions of seabirds that nested on the cliffs were easy prey.

Erik and his men spent the first winter on a small island that he named Eriksey. Afterward they built houses at the head of a long inlet that he named Eriksfjord. They spent the three years of exile exploring the coast, noting places that seemed like good sites for future settlements. They saw no sign of human inhabitants. The winters were long and harsh, but not unendurable, and at the end of three years Erik set out for Iceland in high spirits, sure that many settlers would follow him back to this new land of his. Given his fondness for naming places after himself, he probably considered calling the new country Eriksland. But, as a saga called the *Lándnamabók* (The Taking of the Land) explains, "he believed that more people would go thither if the country had a beautiful name." So, in one of history's first real-estate promotions, he named the country Greenland—although very little of it is ever green—and it has been called Greenland ever since.

In 986 Erik sailed from Iceland to Greenland with a fleet of ships and hundreds of colonists. Under Erik's leadership they established several communities on Greenland's west coast. The eventual fate of these colonies is a tragic and mysterious saga in itself. They flourished for a few centuries, complete with bishops sent by the Pope in Rome and regular trade with Norway and Denmark. Then the North Atlantic climate grew colder, and at the same time Scandinavia was ravaged by several dire epidemics of the Black Plague; the people of Scandinavia had little thought to spare for their kinfolk in faraway Greenland. At the same time, as Europe's trade with Africa and Asia began to increase during the late Middle Ages, furs from Siberia and elephant ivory from Africa replaced the furs and walrus ivory of the Greenland trade. Ships to the Greenland colonies grew fewer and fewer. The last supply run from Norway was made in 1369. Greenland was at first neglected, then all but forgotten.

The Greenlanders were in trouble. Without regular supplies of grain to supplement their diet, they became malnourished. Without the regular arrival of newcomers to the community, they became inbred and weak, and fell victim to disease. And without timber from Norway, they were unable to repair their ships or build new ones, for Greenland has no trees. The Greenlanders' numbers dwindled from a high of about three thousand to a few hundred. The Inuit, the Native Americans of northern Canada, moved into Greenland, clashing with the Norse on many occasions. No one knows for certain what happened to the last Greenlanders. Did they all die, or did some of them join the Inuit? When a ship from Europe finally visited Greenland in the sixteenth century, the visitors found only graves and empty homesteads.

No hint of this bleak future was in the air, however, in the spring of 986, when Erik's followers built their new homes in Greenland. In that same year, the third of the Vikings' three great accidental discoveries took place. This time they found America itself.

The discoverer was Bjarni Herjolfsson, who traded between Norway and Iceland in his knörr. He reached Iceland in the summer of 986, expecting to dispose of his cargo and spend the winter with his father, who lived there. To his surprise, he learned that a few weeks earlier his father had sold the family home and left the island. He had gone off with Erik the Red to live in Greenland. Herjolfsson decided to follow.

Like Naddod and Gunnbjorn Ulfsson before him, Herjolfsson was caught in a gale and blown off course. After the gale subsided, the knörr was surrounded by fog, and when the fog cleared, Herjolfsson was hopelessly lost. Two days later he sighted land, which he described as "level and well-wooded." Before leaving Iceland he had heard enough descriptions of Greenland to know that this was not it. He headed north. Twice more he sighted land—once a flat, rocky shore and once an ice-covered mountain. But these lands were in the west, the wrong direction to be the Greenland settlement, which was on the eastern side of the sea approach. His crew wanted to land and look around, but Herjolfsson refused, saying, "This land looks unwinsome and ungainsome." This is the first recorded opinion about North America—that it looked neither pretty nor profitable.

Herjolfsson decided that he had been blown west past Greenland. He turned east, and four days later he arrived at his father's home near the southwestern tip of Greenland. Herjolfsson settled in Greenland, and it seems that he had no desire to explore the western lands he had sighted. Some Greenlanders, however, *were* interested in his story, perhaps because he reported seeing trees, and wood was always in short supply in the colony. Just as Erik the Red had exploited Ulfsson's accidental discovery of Greenland, so also did his son Leif decide to investigate this new find of Herjolfsson's. Around 1001, Leif Eriksson bought Herjolfsson's knörr and sailed it west with thirty-five men.

Eriksson and his men were the first Europeans to walk the soil of North America. They landed in three places, which were later described in *Erik the Red's Saga*. The first they called Helluland, which means "land of flat stones." It was cold and bleak, and most scholars believe it was either the southern coast of Baffin Island, a large island that lies between Greenland and Canada, or the northern part of Canada's Labrador peninsula. South of Helluland the Vikings made their second landing, at a place they called Markland, or "wooded

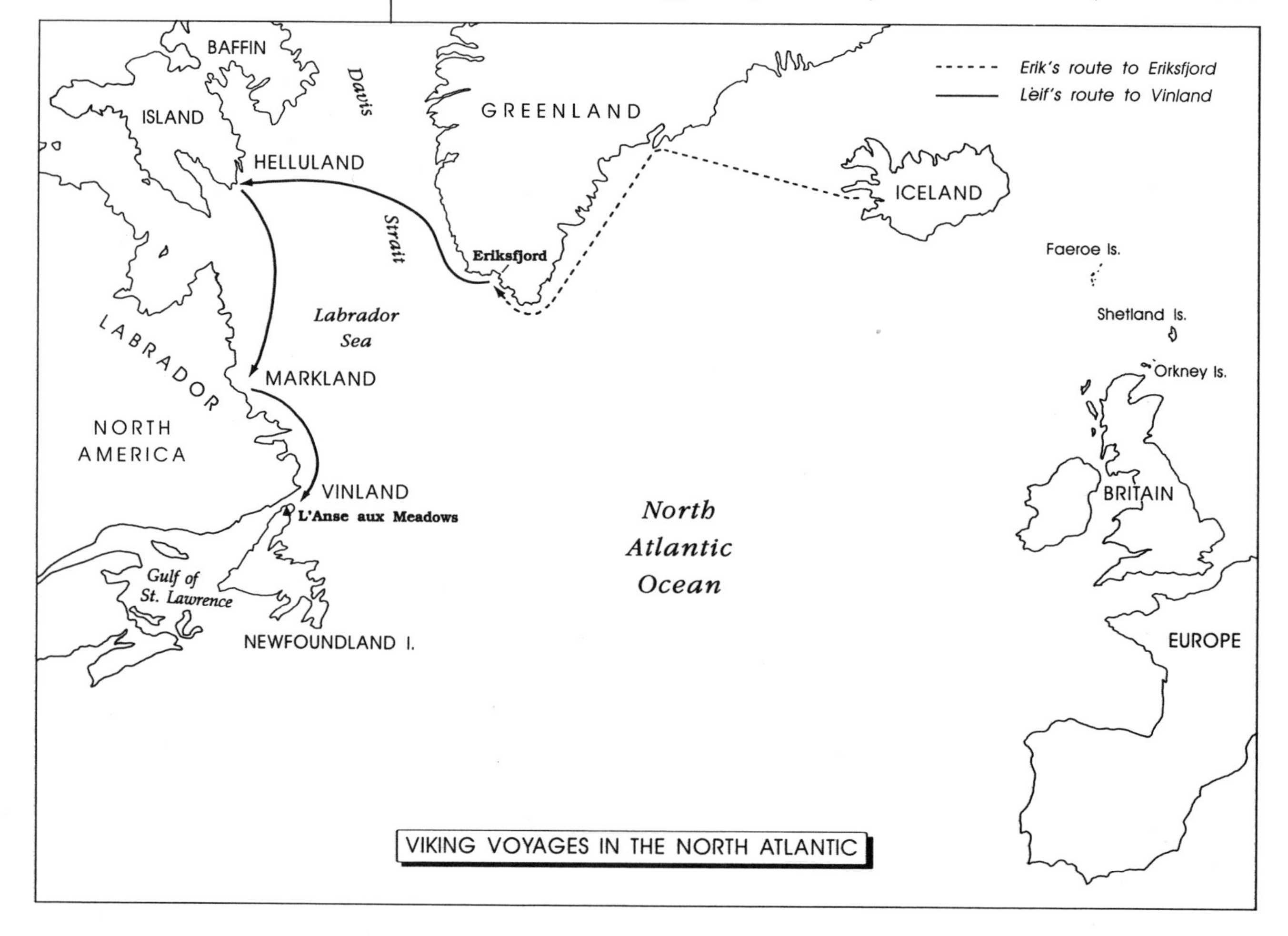

VIKING VOYAGES IN THE NORTH ATLANTIC

land"; this was probably the southern Labrador coast. The third landfall they called Vinland, which could mean "wine land," "vine land," or "fruitful land."

Modern geographers and historians have argued for years about the location of Vinland. Some placed it in Canada, on the shores of the Gulf of St. Lawrence. Others thought it was in Maine, in Massachusetts, in Rhode Island, or even as far south as North Carolina. But most experts now agree that Vinland was probably the northern tip of Newfoundland Island. There, in 1961, at a place called L'Anse aux Meadows, archaeologists uncovered the nine-hundred-year-old remains of a small Viking settlement. Perhaps those ruins are all that remains of the huts Leif Eriksson and his men built during their visit to America.

In the 1960s archaeologists found traces of a Viking settlement on the northern tip of Newfoundland Island, Canada. Half a dozen buildings of timber and sod were built there in the early eleventh century. This modern reproduction was created nearby as a national historic park.

Leif Eriksson and his party had sailed from Greenland in the summer. They explored the three landfalls during late summer and fall, and they spent the winter in Vinland, returning to Greenland the following summer. Leif never went back to Vinland, but his brother Thorvald led an expedition there in 1002 and stayed in the huts Leif had built. This time the Norsemen encountered native inhabitants, either Inuit or forest Indians, and fought with them. Thorvald Eriksson was killed in the fight, and his men retreated to Greenland.

In the years that followed, other relatives of Leif Eriksson tried twice to establish permanent settlements in Vinland. During the first sojourn, a woman named Gudrid gave birth to a boy who was named Snorri. He was the first child of European ancestry born in America. The Norse bartered with the local Native Americans for furs and meat, but fighting broke out again, and the Norse were driven back to Greenland. The final attempt to settle Vinland was led by Freydis, Leif Eriksson's half-sister. It ended in a bloody battle among the settlers, and once again the survivors retreated to Greenland.

That was the end of the Vikings in America, although a few later captains may have landed briefly in Vinland or elsewhere to take on water or cut timber. As the Greenland colony declined and fell out of touch with the rest of the world, Vinland receded into the misty borderland between history and legend. But in the centuries that followed, it was not utterly forgotten. Tales of Leif Eriksson and the western land continued to be told in Iceland, England, and northern Europe, where they may have come to the attention of later explorers—perhaps even Christopher Columbus himself.

Polonia
Asia minor
Arabia
Germania
Hispania
Affrica
Oceanus Occiduus
Cuba
Spagnolla
Brasilia
America
Peru
Mexicum
Circumferentia Centri terræ
Circumferentia Centri magnitudinis
Circumferentia Centri grauitatis
rante de nauios para las Indias.

CHAPTER 2

The Stubborn Quest of Christopher Columbus

"In fourteen hundred and ninety-two,
Columbus sailed the ocean blue."

Generations of American students learned this rhyme from their schoolbooks. Although Christopher Columbus *did* sail the "ocean blue" in 1492, some other often-repeated statements about him are myths.It is sometimes said that Queen Isabella of Spain had to pawn her jewels to pay for his voyage: not true. It has also been said that at the time of Columbus's voyage people believed the earth to be flat, and that his sailors were afraid they would fall off the edge of the world: again, not true. But one thing *is* true: No single event in all of the history of exploration has had such enormous consequences as Columbus's venture. And no other explorer is as controversial as Columbus has become during the last part of the twentieth century. People are now taking a hard new look at what really happened when the two worlds encountered each other.

In a 1621 book on navigation, Columbus appears dressed as a wizard or magician (opposite). Europe and Africa are at the top of the circular map; North and South America are at the bottom. Hispaniola, the island where Columbus founded a colony, is called Spagnolla.

No picture of Columbus drawn from life survives, but each age has created its own image of the explorer. In this 1595 portrait by Theodore de Bry he appears prosperous and successful.

Unlike the journeys of the Vikings, which were few in number and soon almost forgotten, Columbus's voyage to America sparked a bonfire of interest throughout Europe and ushered in a new era of global activity. Mariners, conquistadors, missionaries, and colonists from many nations followed Columbus west across the Atlantic Ocean. Overcoming countless obstacles, they developed new trade routes, built cities, and laid the foundations for dozens of new countries in North, Central, and South America.

But in their ferocious search for gold to send home to Europe, land to claim for the kings and queens of Europe, workers to till the plantations on these new lands, and souls to save for the Roman Catholic Church, the European newcomers killed hundreds of thousands of Native Americans, destroyed whole cultures, and extended the African slave trade halfway around the world. The visits of the Vikings had little or no lasting effect on world history, but Columbus's voyage has left a legacy of mingled bitterness and triumph. In one way, however, Columbus was just like the Vikings: both came upon America entirely by accident.

Columbus was born in about 1451 in Genoa, an outward-looking Italian city that produced many master shipbuilders, mariners, and mapmakers. He went to sea at an early age, probably fourteen, and he made a number of voyages in the Mediterranean Sea. His life reached a turning point in 1476, when he was shipwrecked off the coast of Portugal. Legend says that he floated to shore by holding onto a wooden oar. He ended up in Lisbon, the capital of Portugal, where his brother Bartholomew was in business as a mapmaker.

At that time, Portugal was the center of the seafaring world. Portuguese navigators were making voyages of exploration down the coast of Africa, and Portuguese merchant ships called at all the seaports of Europe. While working in the Portuguese merchant fleet, Columbus made voyages to England and Ireland, and possibly to Iceland as well; he also visited the Azores Islands, a Portuguese colony in the Atlantic Ocean, some 800 miles (1,280 kilometers) west of Portugal—farther west than Iceland. At the time of Columbus's visit, the Azores were considered by the people of southern Europe to be the westernmost land in the world.

Columbus married a Portuguese noblewoman and lived for a time on her family estate in the Madeira island group off the coast of Morocco. The Madeiras were another Portuguese colony in the

Atlantic, south of the Azores and closer to the mainland. They served as a stopping point for ships on the Guinea run, traveling to and from Portugal's outposts along the Gulf of Guinea in West Africa. Columbus may have made at least one voyage to Guinea. He certainly talked with his fellow mariners and absorbed every scrap of geographical and navigational information he could find. During these years, he developed a new theory about the location of the Indies.

Everyone in Europe at that time wanted to get to "the Indies"—an all-purpose term for southern Asia that included India itself and also the little-known lands east and south of India. China, which had been visited by Marco Polo and other Europeans in the thirteenth century, was known to be in the general region of the Indies as well. Since ancient times, the Indies had been the source of goods highly prized in Europe: silks and other fine textiles, precious gemstones, and, above all, pungent and fragrant spices such as pepper, nutmeg,

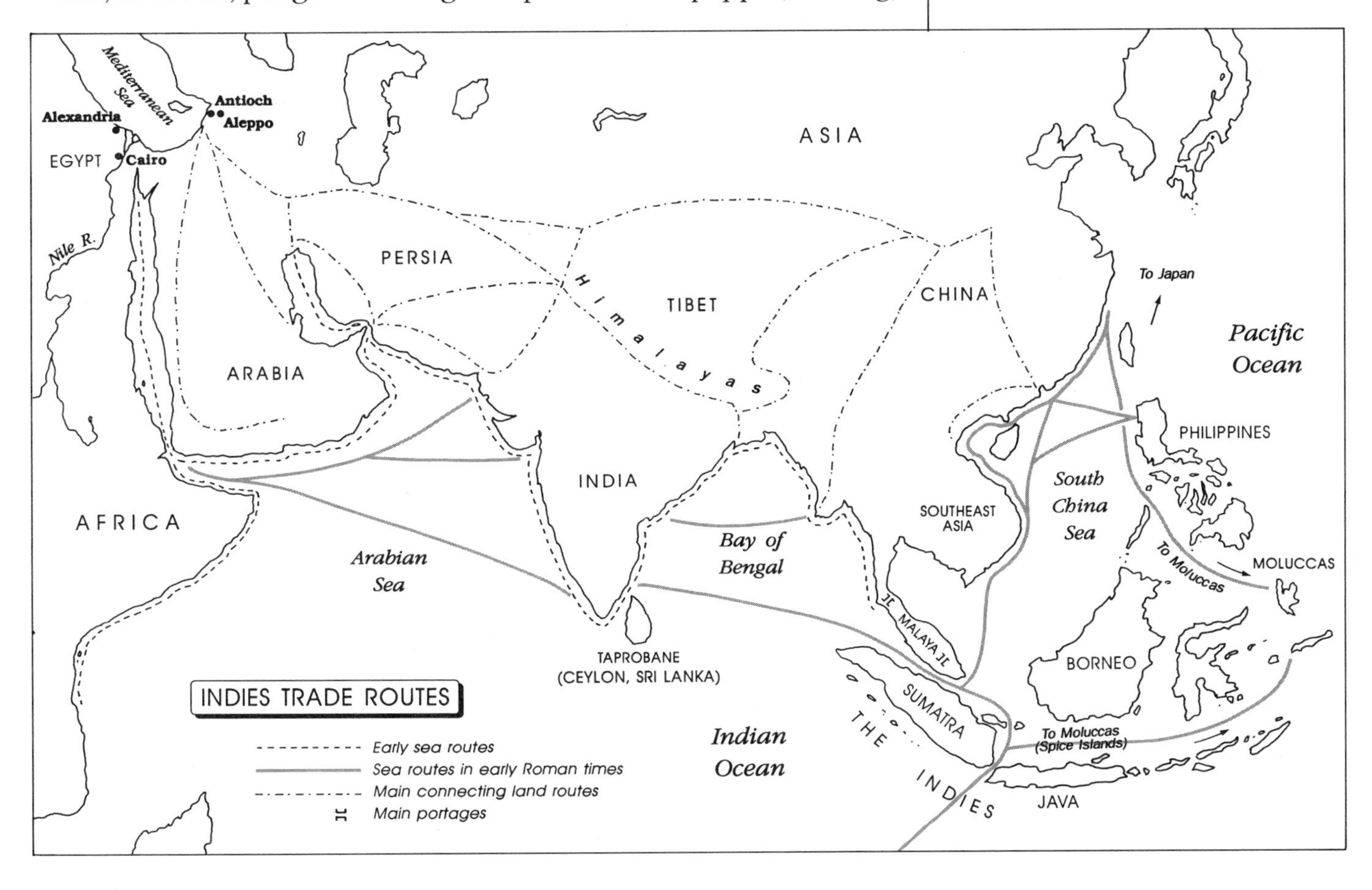

cinnamon, and cloves, which were used both to preserve food, especially pickled meat, and to improve its taste.

Unfortunately, Europe's ability to obtain these desirable items was painfully limited. The rise of the Islamic Empire in Asia had closed off the caravan routes to the east. Goods from the Indies reached Europe by sea, and only after passing through many hands. Chinese and Malay traders sold them to Hindu traders in India, who sold them to seagoing Arab merchants, who shipped them across the Indian Ocean to ports in Egypt or the Middle East. Goods from the Indies wound up in Arab trading capitals such as Alexandia, Egypt, or Aleppo, Syria, where European merchants could buy them. But each transaction raised the price, making pepper, for example, one of the costliest commodities in the world.

By the fifteenth century, the nations of Western Europe were looking for a way to reach the Indies directly, bypassing the intricate and costly Muslim trade monopoly. Christopher Columbus became convinced that he had discovered the way. He believed he could reach the easternmost part of Asia by sailing *west* from Europe.

Educated people of Columbus's time were aware that the earth is a sphere. With a little prodding, they could picture the lands and seas curving around to meet each other like a map wrapped around an orange. The first known globe, in fact, was made in 1492, the year of Columbus's historic voyage, by Martin Behaim, a German navigator and geographer who lived and worked in Portugal. The globe does not show Columbus's discoveries; it was completed before they were known. But its very existence does show that many people were familiar with the idea of a round world.

A woodcut made in Spain in 1496 shows Columbus setting out for the Indies in his flagship, the *Santa María*.

Most people could see that there was nothing really wrong with the idea of sailing west to get to the Far East. The real questions were: How great was the distance? How long would the voyage take?

Columbus worked out a set of extremely complicated calculations, coming up with a distance of 2,400 miles (3,840 kilometers) from the Canary Islands to Japan, which no European had ever visited but which Marco Polo had mentioned in his account of his travels in China. Columbus's estimate did not sound too fearful—after all, the Portuguese had made successful voyages of almost 5,000 miles (8,000 kilometers) along the African coast.

But Columbus's calculations were wrong. He had greatly underestimated the size of the earth, and he had greatly overestimated the

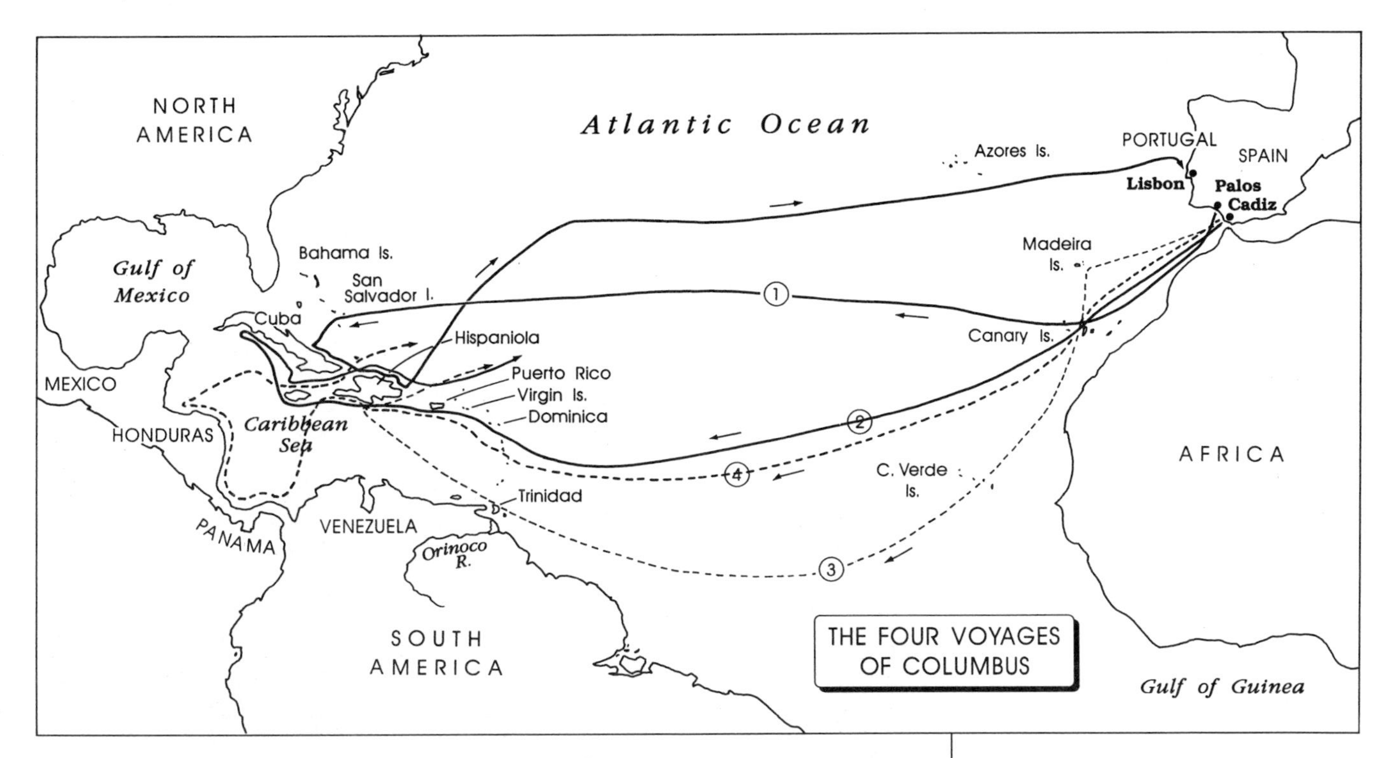

eastward extent of Asia. The real distance from the Canaries to Japan is 10,600 miles (16,960 kilometers)—almost four and a half times farther than Columbus thought. Of course, no one knew that two unknown continents *and* an unknown ocean lay in the way.

Columbus called his scheme for reaching Asia the Enterprise of the Indies. In 1484 he presented the idea to King João II of Portugal and asked João to provide ships and men for the voyage. But according to Portuguese court historian João de Barros, Columbus came off as "a big talker, and boastful," and the king "gave him small credit." Rejected by Portugal, Columbus spent eight years trying to persuade England, France, or Spain to back his venture. Finally, in 1492, King Ferdinand and Queen Isabella of Spain agreed to sponsor Columbus, and the Enterprise of the Indies was under way.

Columbus's little fleet of three ships left the Canary Islands on September 6, 1492. With every expectation of finding himself in Marco Polo's Asia, Columbus carried royal letters of introduction to the Great Khan, or Emperor of China. On October 12 the fleet fetched up against a low-lying green island in a warm blue sea.

Modern historians have argued about the exact site of Columbus's first landfall even more than they have argued about the location of Vinland. But since the 1980s most have agreed that Columbus first set foot in the Americas somewhere in the Bahama Islands, probably on either Watlings Island or Samana Cay. The Native Americans who lived there called the place Guanahani; Columbus called it San Salvador. He thought that it was one of the many islands said by Marco Polo to dot the seas of the Indies, so he called its inhabitants Indians.

Columbus's fleet felt its way through the Caribbean Sea, making several more landfalls in the Bahamas and then sailing south and

AMERICAE
PARS QVARTA.
Sive,
Inſignis & Admiranda Hiſtoria de reperta
primùm Occidentali India à Chriſtophoro
Columbo Anno M. CCCCXCII
Scripta ab Hieronymo Bezono Mediolanenſe,
qui iſtic ānis XIIII. verſatus, diligēter omnia obſerva-
vit.
Addita ad ſingula ferè capita, non contemnenda ſcholia
in quibus agitur de earum etiam gentium idololatria.
Acceſsit præterea illarum Regionum Tabula
chorographica.
*Omnia elegantibus figuris in aes inciſis expre-
ſsa à Theodoro de Bry Leodienſe, cive
Francofurtenſi Anno cIↄ Iↄ XCIIII.*
Cum prevelegio S. C. Maieſtat.

An account of the 1492 voyage that was published in 1594 featured pictures of the people of the Americas. Columbus had called them Indians because he believed he was in the Indies.

east along the coasts of Cuba and Hispaniola (the island now shared by Haiti and the Dominican Republic). The flagship, the *Santa María,* was wrecked on a coral reef off Hispaniola, and Columbus transferred his command to the smaller *Niña.* All the while, he looked in vain for the gold-roofed palaces he expected to see in Japan, or for the large, ancient, bustling ports that he knew lined the coast of China. His search was fruitless. On January 16, 1493, he headed back to Spain, convinced that he had landed in a previously unknown part of the Indies, probably the easternmost fringe of Asia.

He was greeted with acclaim and delight by the king and queen of Spain, who granted him the title Admiral of the Ocean Sea and speedily fixed him up with seventeen ships for a second expedition. He set out again in September 1493, carrying twelve hundred men to start a colony on Hispaniola. In Columbus's eyes, the colony was a trading post and shipping port for the booming trade he expected to start with the merchants of Asia; in reality it was the first permanent European settlement in the Americas.

After setting up his colony, Columbus took three ships to explore the coast of Cuba, which he thought was part of the Asian mainland. Doubt must have gnawed at him, though, or perhaps some of his crew questioned his claim that these bleak jungle islands were in fact the golden Indies of legend. He ordered the entire crew to sign an oath that they believed Cuba to belong to the continent of Asia. The oath also declared that Columbus would certainly have encountered "civilized people of intelligence who know the world" if he had only sailed on a little farther. Everyone signed—Columbus had threatened to cut out the tongue of anyone who refused.

After leaving Cuba, Columbus landed at Jamaica and Puerto Rico. He had hoped to find the sophisticated civilizations of the Indies, but instead he found the Native American Taino and Arawak peoples who, in his eyes, were nothing more than naked savages. Columbus claimed these lands for Spain and went on looking for someone to direct him to the Indies. Finally he gave up, and in 1496 he went back to Spain to report to his royal sponsors.

This time the reception was much less enthusiastic. King Ferdinand and Queen Isabella had invested an enormous amount of money in the Enterprise of the Indies, and so far they had received no spices, no gems, and no letters of friendship from the Great Khan—only a few crude golden trinkets and the claim to some undeveloped real estate. Furthermore, people were beginning to wonder whether Columbus really had reached the Indies.

Columbus, however, refused to consider the possibility that his theory was wrong. Stubbornly he mustered evidence to support his claim. He said that a nut-bearing tree he had found in Cuba was the Asian coconut palm described by Marco Polo; it was not. He said that some of the native people he met in Cuba on his first voyage had referred to the Great Khan with the words "El Gran Can"—which, if true, would mean that the Taino people spoke Spanish before

Columbus's arrival, a startling development indeed. What the Indians *had* said was "Cubanacan," the Taino word for a place in central Cuba.

It took Columbus several years to persuade Ferdinand and Isabella to outfit a third voyage. In the meantime, other expeditions set out—expeditions that would ultimately disprove Columbus's claim. But in 1498 Columbus doggedly went west a third time, and on this voyage he *did* discover a continental mainland. It was the coast of South America.

Columbus saw the outpouring of Venezuela's Orinoco River and realized that such a large river must come from the interior of a continent, not an island. Turning to biblical geography to help him make sense of the obstacles he kept encountering on his way to Japan, he claimed that he had found "the spot of the earthly paradise," a sacred and hitherto unknown land that was the setting of the Garden of Eden and the source of the four rivers of Paradise mentioned in the Bible. Yet still he believed he could sail around Paradise to the Indies. He retreated to Hispaniola to plan his next attempt.

Columbus and his brothers Bartholomew and Diego, who had been placed in charge of the colony on Hispaniola, were poor administrators. Beset by rebellion among the settlers and hostility from the natives, they badly mismanaged the colony. Reports of trouble in Hispaniola reached Ferdinand and Isabella in letters sent back to Spain on supply ships. They sent an officer named Francisco de Bobadilla to take charge of the colony, and Bobadilla had Columbus arrested and shipped back to Spain in 1500. But Columbus remained as fanatically stubborn as ever; perhaps by this time he was even a little crazy. Determined to prove he had reached the Indies, he petitioned the king and queen for still another chance.

The islanders who greeted Columbus on Hispaniola were friendly and generous, but they did not rush to give him elaborate jewelry and other European-style treasures. This sixteenth-century illustration owes more to fantasy than to fact.

They agreed, although they forbade Columbus to return to Hispaniola, and he set out in 1502 with four ships on what he called *El Alto Viaje* (The High Voyage). He carried a letter to the Portuguese navigator Vasco da Gama, who sailed for India by another route—east around Africa—around the same time. The letter was never delivered, because, of course, Columbus never reached India.

He sailed along most of the Central American coast, expecting that at any minute a bay or strait would open up and let him enter the Indian Ocean. But the coast remained stubbornly impenetrable, and the golden temples and rich ports of the Indies continued to elude Columbus. He returned to Spain for the last time in 1504 and died two years later, bitterly disappointed at his failure to find Marco Polo's fabulously wealthy Oriental kingdoms—but still unshakably convinced that he had sailed along the edge of Asia.

"Columbus was wrong, but lucky," says science historian John Noble Wilford. "No explorer succeeds without some luck." Columbus's luck probably saved his life, for it is unlikely that he and his crew could have survived the long, long voyage more than halfway around the globe to Asia if the Americas had not blocked the way.

Yet Columbus was also singularly unlucky. He had doubled the size of the world known to the Europeans of his day, but he did not rejoice in this achievement because he had failed at what he had originally set out to do: He had not been able to find Japan, China, or India. And, although his voyage launched an era of exploration and colonization in what was to Europeans a New World, the new continents he found do not bear his name. Instead, they were named for another explorer, an Italian banker and merchant who explored the coast of Brazil on behalf of King Manoel I of Portugal in 1501 and 1503. This voyager's name was Amerigo Vespucci.

The Spanish government shipped Columbus back to Spain in disgrace in 1500. Nevertheless, he made one more voyage to the New World, still searching for the passage to the Indies.

Vespucci's letters describing his journeys were passed from hand to hand and read throughout Europe. In one of them he says that the land he explored was truly a *Mundus Novus*, the Latin for "New World." This is the first known description of the western lands as new continents, not as part of Asia. One of the people who read Amerigo Vespucci's letters was a German mapmaker named Martin Waldseemüller, and in 1507 he placed the name "America" on one of his maps, approximately where Brazil is today. He said, "I do not see why anyone should object to its being called after Americus the discoverer." ("Americus" was the Latin version of "Amerigo.") Later Waldseemüller tried to drop the name, but it had already been adopted by other mapmakers, and within a few years the western continents were being called North America and South America.

The Americas were discovered by accident, and their naming was equally haphazard. It is by mere chance—Christopher Columbus's final piece of bad luck—that they are not called North Columbia and South Columbia.

17

VARES·CABRAL·ANO·DE·500·

CHAPTER 3

Cabral Bumps into Brazil

One of the reasons that King João of Portugal turned down Christopher Columbus's Enterprise of the Indies in 1484 was that Portugal had spent many years, much money, and the lives of scores of mariners exploring its own route to the Indies. The Portuguese route went east instead of west. Portugal planned to reach India by sailing down around the great bulk of Africa and then up into the Indian Ocean. In 1498, while Columbus was probing the coast of Venezuela in the vain hope of finding a passage to the Indies, Vasco da Gama was making the first trip from Portugal to India by this eastern route.

No sooner had Gama returned from India than the king of Portugal launched a large fleet—thirteen ships and twelve hundred men—under the command of Pedro Álvars Cabral. The fleet set sail in 1500. Cabral's mission was to follow Gama's route to India, make profitable trade arrangements with the rulers and merchants there, and come back to Portugal with the holds of his ships bulging with spices, gems, and silk. But Cabral owes his place in history to something that was not planned at all: the accidental discovery of Brazil.

How could a ship sailing from Portugal to India accidentally bump into Brazil, which lies in the opposite direction halfway around the world from India? At first glance it seems that Cabral must have been hopelessly lost, blundering around the Atlantic, in order to have reached Brazil when he was aiming for the southern point of Africa.

Pedro Álvars Cabral was not supposed to be an explorer. His Portuguese fleet (opposite) was supposed to bring home a wealth of gems, spices, and silk from the east, not discover new lands in the west.

Vasco da Gama pioneered the Portuguese sea route around Africa to India. It was on his advice that Cabral sailed far out into the Atlantic Ocean—and unexpectedly landed in Brazil.

Yet in reality he was following good sailing practice—he was going in the right direction, but he went a little too far.

The Atlantic Ocean contains two immense circular patterns of wind and water currents called gyres. The North Atlantic gyre flows clockwise, down along the coast of Europe, west across the North Atlantic Ocean just above the equator, and up along the coast of North America. The part of the gyre that flows from south to north along North America and then east across the top of the North Atlantic is called the Gulf Stream.

The South Atlantic gyre flows counterclockwise, up from south to north along the African coast, west across the South Atlantic just below the equator, south along the coast of South America, and finally east toward southern Africa. Mariners before Cabral had encountered endless difficulties trying to fight their way south along the African coast against the wind and water currents that steadily pushed north.

Drawing upon the experience of several generations of sea captains who had explored the African coast, Gama advised Cabral that the best way to get to the bottom of Africa was not to aim straight for it but to sail out into the Atlantic in a wide arc to the southwest so that he would be carried along on the gyre. Gama himself had rounded Africa this way, and, although he did not realize it, he came within 600 miles (960 kilometers) of South America. Cabral decided to follow his advice. It was good advice; sailing vessels still follow Gama's route today.

Less than two months after leaving Portugal, Cabral was well south of the equator. But his fleet had ridden the currents of water and wind farther west than Gama's had done. On April 22, 1500, a lookout high in the mast of Cabral's flagship sighted a mass of floating vegetation in the water ahead—a sign of land nearby. A few hours later, the cry went up: *"Tierra! Tierra!"* Land had been spotted. Cabral put in to shore to investigate.

Like all good explorers in the age of empire building, Cabral claimed the new land for his monarch, calling it Tierra da Vera Cruz (Land of the True Cross). He spent less than two weeks there and explored less than 50 miles (80 kilometers) of coastline. He discovered little that he thought was interesting: some friendly natives, but no cities, gold, or spices. So Cabral put up a wooden cross to mark his claim, had his shipboard priest say Mass on the shore, sent one vessel from the fleet back to Portugal to report his unexpected find, and then calmly resumed his interrupted journey to India.

He left more behind in Tierra da Vera Cruz than a wooden cross, however. His fleet, like most expeditions of his era, carried a dozen or more convicts who had been sentenced to death at home. It was to these members of the expedition that all the disagreeable, dangerous, or potentially fatal tasks were assigned. Cabral chose two of them to remain behind in the newly discovered country. They were supposed to learn the local language, befriend the natives, and gather information about the prospects for trade or treasure. With luck, some future expedition would pick them up, and then their information would be useful.

Perhaps the two convicts ought to have been happy to leave the stinking, crowded ships and have a whole new country to explore. Yet they are said to have wept as the fleet pulled away. And it is possible that their death sentence was carried out after all, in a gruesome way; later travelers learned that the Tupinamba, the natives who inhabited the region, were cannibals.

Cabral did reach India, and he did return to Portugal with cargoes of spices and gems. But he lost six of his ships and about three-quarters of his men, and the fights he had with Indian traders and

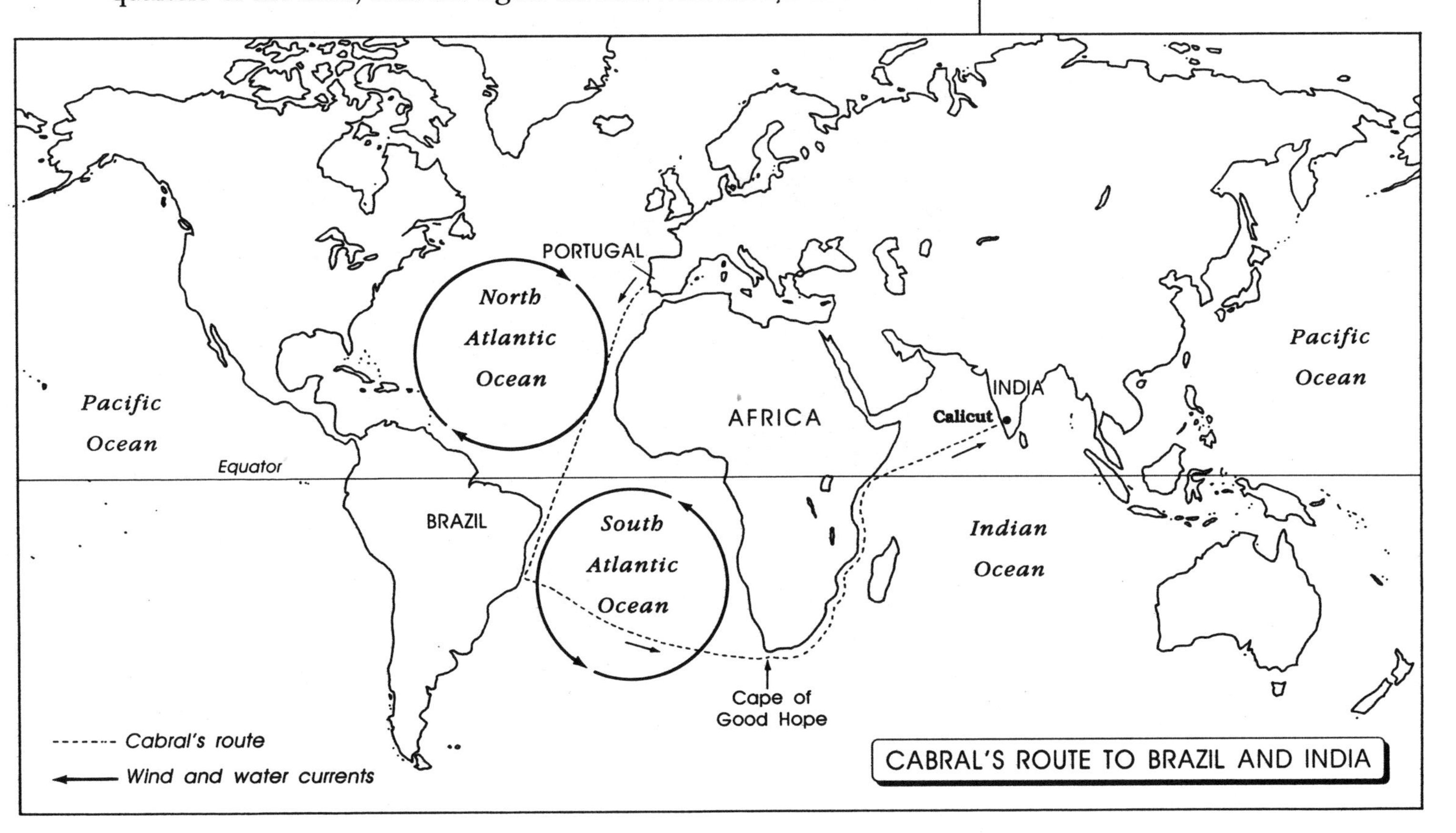

CABRAL'S ROUTE TO BRAZIL AND INDIA

While in Brazil, Cabral held a Catholic Mass for his crew. Although the friendly Indians of the region certainly attended, it is unlikely that they prayed in the manner shown in this early depiction of the scene.

rulers made life very difficult for the next Europeans who reached India. So Cabral's mission to India was only a partial success. It has been largely forgotten by history. Instead, Cabral is honored throughout Brazil as the European discoverer of that country.

Should Cabral be celebrated as Brazil's discoverer? As Anthony Smith points out in *Explorers of the Amazon,* four Spanish expeditions touched the coastline of South America in 1499, a year before Cabral's voyage, and three of them visited the northern edge of what is now Brazil. Columbus himself had landed in Venezuela in 1498. But all these landings were part of the explorations Columbus had started in the Caribbean. In contrast, Cabral found land where no one had expected it. His landfall at Monte Pascal, north of present-day Rio de Janeiro, was the southernmost point in the Americas yet reached by a white man. But the main reason Cabral is called the discoverer of Brazil is that Brazil became a Portuguese colony, and the Portuguese people of Brazil wanted their hero to be a Portuguese, not one of the rival Spanish. Cabral died in obscurity, without ever returning to the place he named Tierra da Vera Cruz. Even the name he chose was later discarded. Yet Brazil remained Portugal's largest overseas colony, ninety-six times the size of the parent country, until it achieved independence in 1822—and all because Cabral sailed a little too far west.

Cabral is regarded as the discoverer of Brazil. His brief visit laid the foundation for Portugal's claim to its largest colony.

CHAPTER 4

Down the Amazon by Mistake

The Spanish conquistador Francisco de Orellana was the first European to explore the Amazon River.

Once Columbus, Cabral, and other explorers had pioneered the way, a horde of rapacious *conquistadors* (Spanish for "conquerors") descended upon the Americas. These conquistadors were Spanish warlords determined to seize the riches of the New World—and, if necessary, to destroy whole Native American civilizations to do so. Hernán Cortés overcame the Aztec Empire in Mexico in 1521, and Francisco Pizarro conquered the Incas of Peru between 1532 and 1533.

The conquistadors were not idealistic explorers, drawn by the lure of the unknown. They wanted gold and silver first, land and slaves second, but they had no love of adventure for its own sake. Nevertheless, one of them did have an extraordinary adventure, quite by accident. Without knowing what he was doing, he became the first European to travel the full length of the world's largest river.

Francisco de Orellana was one of Pizarro's lieutenants in the conquest of Peru. He founded the city of Guayaquil, on the coast of present-day Ecuador, where he settled down to live on an estate granted by the Spanish crown as a reward for his services.

In 1541 Orellana heard of an expedition into the interior of the continent that was being planned by Gonzalo Pizarro, Francisco's

An Amazon, one of the mythical women warriors for whom the world's largest river was named (opposite). The first Europeans who explored the river saw fighting women and were reminded of the legendary Amazons.

half-brother. From Quito, a city in the interior of Ecuador, on the western slope of the Andes Mountains, Gonzalo Pizarro intended to cross the mountains and plunge into the unexplored jungle on the other side. He was searching for gold and also for cinnamon, which was rumored to grow east of the mountains. Cinnamon was worth almost as much as gold back home in Europe.

Orellana decided to join Pizarro's expedition, which consisted of between two hundred and three hundred Spaniards and more than four thousand Indian slaves. With twenty-three soldiers from among his own followers, Orellana made his way by narrow trails through the towering, snow-capped mountains. He caught up with Pizarro 110 miles (176 kilometers) east of Quito and was made Pizarro's second in command.

By this time, however, months had passed since Pizarro had left Quito. Food was running low, and the hoped-for golden treasures and forests of cinnamon trees had failed to materialize. Progress was agonizingly slow and difficult. The men in their heavy armor had to contend with the tropical forest, which was damp, marshy, hot, mosquito-infested, and threaded with countless streams that had to be bridged or waded. It rained every day. Many of the men became ill with fever or dysentery.

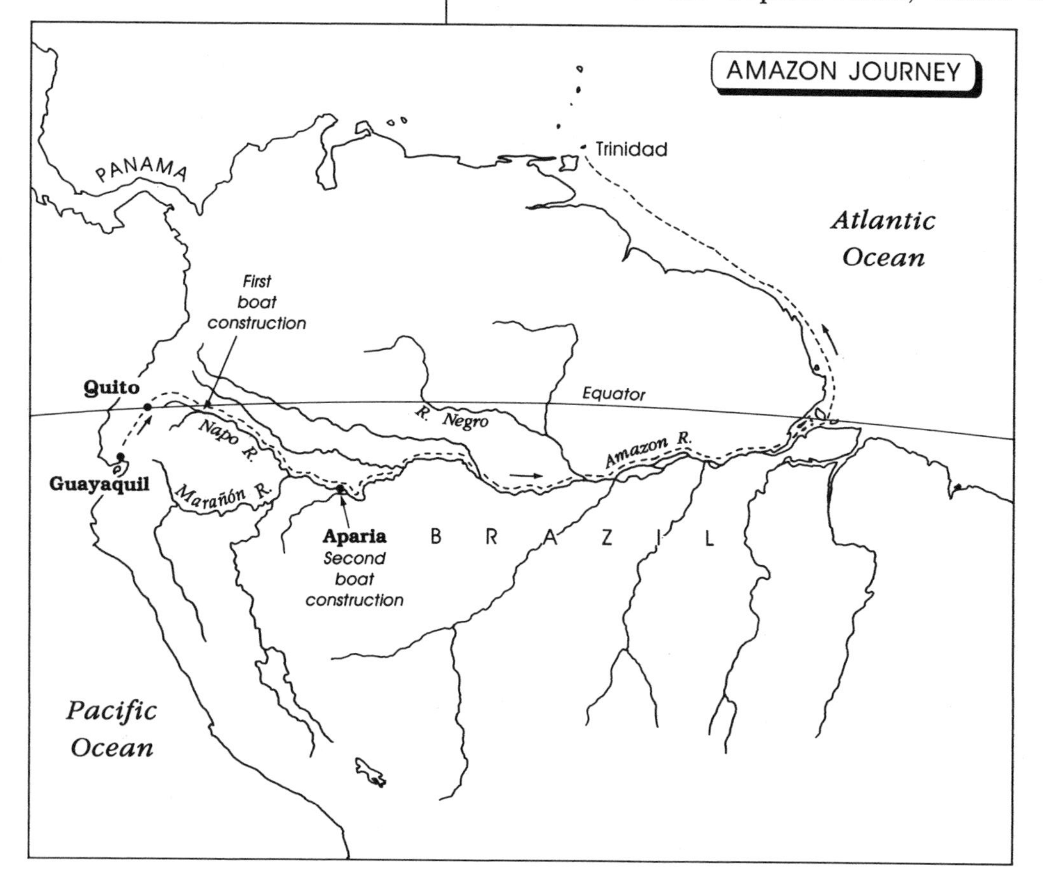

The travelers entered a region of many rivers. These waterways provided the only highways through the trackless forest, and the Spanish decided to follow a good-sized river east. They were able to obtain some Indian canoes, by purchase or by force, and they also built a sturdy wooden boat. They struggled on, some on the river and the others slogging with their horses along the riverbank. They found few Indians and no food.

On Christmas day, December 25, ten months after leaving Quito, Pizarro and his men were in bad shape. They were weak, ill, and in desperate need of food. It was decided that Orellana would take 57 of the men and hurry ahead in the boat and the canoes. No one knew what lay downstream, but Orellana expected to find Indian villages where he could obtain provisions. He and his comrades promised to return as soon as they found food. Pizarro and about 140 men watched them go with hope and hunger in their eyes.

The river Orellana set out on was the Napo River. It flowed along quite swiftly—so swiftly that Orellana and his companions must soon have realized how difficult it was going to be to paddle back upstream. After a week or so they came upon their first Indian village. Leaping ashore, they captured the village and its supplies of food—mostly wild turkeys, fish, and fruit. No Indians were harmed, and Orellana took pains to reassure and befriend the natives. The Spaniards remained in the village for a time, puzzling over what they should do next.

Their duty was to return upstream to Pizarro and the others, as they had promised: anything else was treachery. But the fast-flowing Napo had carried them about 700 miles (1,120 kilometers) from Pizarro's camp, and they knew they would find no food on the way back. And why should they go back? They could do no possible good. By now, Orellana and his men reasoned, Pizarro and the others had either died or moved on. Perhaps they had given up and gone back to Quito.

The strongest argument of all was the river itself. Orellana and his men did not believe they could paddle fast enough or hard enough to retrace their course against the current; and as for going back by land, it would have taken them more than a year to hack their way through the jungle. So they decided to go on. They thought that if they stayed on the river they might come out somewhere in the Caribbean Sea, where there were plenty of Spanish bases. They left the village, heading east, on February 2, 1542.

Meanwhile, back at Pizarro's camp, the expedition's commander fumed with anger. He and his men were reduced to eating their horses and dogs. Days passed, then weeks, and Orellana did not return. In Pizarro's eyes, there was only one explanation: Orellana, as he complained later in a letter to King Charles V of Spain, had "gone

The Spanish built several boats, melting their bullets to make nails. Although this illustration of the boatbuilders shows two men using a saw to cut timber, the Spanish probably had only swords and axes to use as tools.

off and become a rebel." Pizarro raged about his lieutenant's desertion, but Orellana was beyond his reach. Vowing to take revenge on the deserter, Pizarro led a retreat back to the Pacific coast.

It was a long and terrible journey. The men suffered from heat and fever in the jungle, from bitter cold in the mountains, and from starvation all the way. In August 1542 Pizarro and the ragged, skeletal survivors staggered into Quito. Pizarro then wrote the letter to King Charles, in which he accused Orellana of "displaying toward the whole expeditionary force the greatest cruelty that ever faithless men have shown." Historians ever since have debated whether Orellana maliciously abandoned Pizarro to seek his own fortune or whether, having committed himself to the river, Orellana discovered that he could not turn back. The truth will probably never be known.

Little more than a week after Orellana and his fellows had resumed their journey down the Napo, they found that it flowed into a much larger river. Their little boats were swept onto a realm of water so wide that they called it an "inland sea." They did not know it, but they were now on the largest river in the world. (The Nile, in Africa, is longer, but the South American river is much wider and carries more water: 20 percent of all the world's river water.) Although the Spaniards did not know it, the river stretched ahead of them for more than 2,000 miles (3,200 kilometers).

The gold-hungry conquistadors feared death at the hands of hostile Indians, who were said to kill the Spanish by pouring molten gold down their throats.

They *did* know that they needed a bigger boat. They managed to build one in April, at a native settlement they called Aparia, where they received food and help from the village chieftain and his people. The rest of their journey was a constant struggle against hostile Indians who came at them in flotillas of huge canoes, shooting arrows and poisoned darts, and against the river itself, with its sudden fierce storms, its turbulent currents, its giant crocodiles and stinging electric eels, and its confusing mazes where the channel wound through scores of desolate islands.

Not all the Indians were hostile; some were quite friendly, though cautious. But one skirmish with a warlike tribe had long-lasting consequences. It took place near the Río Negro, a large tributary of the main river. During the fight, the Spanish saw a dozen or more women among their opponents.

Friar Gaspar de Carvajal, a monk in Orellana's group who wrote the only surviving account of the trip, says, "For we ourselves saw these women, who were there fighting in front of all the Indian men as women captains, and these latter fought so courageously that the Indian men did not dare to turn their backs; anyone who did turn his back they killed with clubs right there before us. This is the reason that the Indians kept up their defense for so long." The Spanish immediately dubbed these women warriors Amazons, after a legendary tribe of fighting women who were believed by the ancient Greeks to live near the Black Sea, on the border between Europe and Asia.

After months of arduous travel through the unknown, in August the voyagers emerged from the wide mouth of the river onto the Atlantic Ocean. But their troubles were not over yet. They had almost no food, and their crude boats were extremely hard to handle in the open sea. Nevertheless, they made their way north along the coast until, after several weeks, they reached a Spanish base on Margarita Island, off the coast of Venezuela. At last their long journey was over.

Orellana and his followers were the first Europeans to cross South America from west to east, and they were also the first to ride the mighty Amazon River across the continent. What had started as a simple foraging mission had turned into an epic river adventure.

Orellana went back to Spain and organized an expedition to explore the vast Amazon River basin. The expedition was poorly funded and badly managed, however; it ended in shipwreck and disaster in the mouth of the Amazon. Orellana died there in 1546

Indians take aim at a tree-climbing creature, probably intended to represent a sloth. Fanciful drawings of the people and wildlife of the Amazon appeared in many European geography and travel books.

at the age of thirty-five. Gonzalo Pizarro was killed in 1548 by a Spanish royal army after trying to set himself up as the king of Peru.

Francisco de Orellana's name and fame outlived him for a while. The Spanish knew about the existence of his river, for some of them had sailed into its mouth. They had originally called it the Marañón, but it was renamed for Orellana after news of his journey spread. For a few years the river was called the Rio de Oregliana, as it appears on Spanish maps of the late sixteenth century. Before long, however, most people had forgotten the details of Orellana's adventure; they remembered only the tantalizing episode of the women warriors whom the priest Carvajal had called Amazons. Travelers and mapmakers began referring to the river as the Rio de Amazonas, and the name stuck. The river is still called the Amazon today. But the women warriors, if they ever existed, were never seen again, although many later travelers to South America looked for them.

PART TWO

IN SEARCH OF A PHANTOM

Columbus found a world he had not sought, and Cabral found a continent where he expected only open sea, but many explorers did just the opposite. They set out in search of something they thought existed and never found it. Much of the history of discovery is about people *not* finding what they were looking for. In the process, though, they sometimes made discoveries that were as interesting as they were unexpected.

IOANES·PRESBR·MAX·DE·IDIA·ET·ETHIOPIA·
·FVGE·SVPERBIAN·TER·
·FVGE·LVXVRIA·DELIGNO·
·FVGE·GVLAM·DEPLVMBO·
·FVGE·IRAM·DE·FERRO·
·FVGE·INVIDIAM·DECVPRO·
·FVGE·ACIDIAM·DEARGENTO·
·FVGE·AVARITIAM·DEAVRO·
PRESTO·GIOVANNI·DE·INDIA·ET·ETHIO

CHAPTER 5

Prester John's Mysterious Kingdom

The legend of Prester John may be the biggest and longest-lasting practical joke in history.

"Prester" is short for *presbyter*, or priest, and Prester John was believed to be a powerful priest-king who ruled over a mighty Christian kingdom somewhere beyond the fringe of the known world. Stories about him began to appear in the twelfth century, a time when the Christian nations of Europe felt very vulnerable indeed. The Islamic Empire had surged out of its birthplace in Arabia to engulf North Africa and all of the Middle East, including the Holy Land of the Christian Bible. Europeans launched the Crusades, a centuries-long series of expeditions intended to win back the Holy Land, but the Muslims—or Saracens, as Europeans of the Middle Ages called the Arabs and other Islamic peoples—were fierce opponents. To the people of Europe, the world seemed increasingly large and hostile, and they longed for an ally.

Miraculously, they found one—or thought they did. In 1122 a traveler reached Rome who claimed to be a Christian bishop from India, where the Apostle St. Thomas was believed to have died. This gave rise to rumors of a Christian kingdom somewhere in India, a land that was almost entirely unknown to Europeans.

Then, in 1145, a German bishop named Otto attended a conference in Italy, and there he met the Bishop of Antioch, Syria

For centuries the legend of Prester John haunted the European imagination. The title page of a 1495 book about Prester John portrays him as the emperor of India and Ethiopia (opposite).

(today Antioch is in Turkey). The Syrian bishop told Otto that a few years earlier a king named John, who lived far in the mysterious East beyond Persia (present-day Iran), had fought a tremendous battle with the Saracens. According to the Syrian, this John was not only a king but a Christian priest, descended from the Three Magi of the Bible, and his realm was an outpost of Christianity. John had tried to come to the aid of the European Christians who were fighting the Saracens in Jerusalem, but the way had been blocked by the Tigris River. John had no means of getting his army across the river unless the water froze. "After halting on its banks for some years in expectation of a frost," the account ran, "he was obliged to return home."

Otto was greatly excited by this news and wrote an account of it that was widely circulated. People started to think of this unknown

A thirteenth-century manuscript shows an army of Saracens, or Muslim fighters, chasing a Crusader knight (with a cross, the emblem of Christianity, on his shoulder).

Prester John as their ally in the East—although no one seems to have asked how helpful an ally he would be if he could not even get an army across a river.

The next news of Prester John reached Europe in 1165 in the form of a ten-page letter, allegedly written by John himself to the three most important men in Christendom: Pope Alexander III, Holy Roman Emperor Frederick Barbarossa, and Emperor Emanuel I of the Eastern, or Byzantine, Roman Empire (based in what is now Turkey). The letter caused a sensation. It was quickly translated into many languages and circulated around Europe.

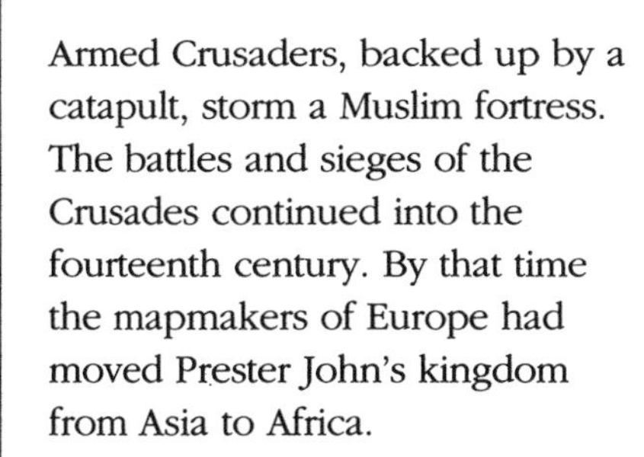

Armed Crusaders, backed up by a catapult, storm a Muslim fortress. The battles and sieges of the Crusades continued into the fourteenth century. By that time the mapmakers of Europe had moved Prester John's kingdom from Asia to Africa.

In the letter Prester John boasted that his kingdom was the richest and most powerful that had ever existed. John ruled over seventy-two kings, and he was so far above these ordinary kings that they cooked his meals and waited on his dinner table. His kingdom contained many wonders, including a river of jewels, a land of dog-headed men, a realm of women warriors, centaurs (beings who were half man, half horse), dragons, a magic mirror, a fountain of youth, the Tower of Babel, and much, much more. "If you can count the stars of the sky and the sands of the sea," wrote John, "you will be able to judge thereby the vastness of our realm and our power."

Pope Alexander wrote back to Prester John. Alas, he was unable to put an address on the letter, because no one knew where Prester John's kingdom was located. So the Pope gave the letter to his doctor, one Magister Philippus, and sent him off to Asia to find Prester John and deliver the letter. Magister Philippus was never heard from again, and the potentates of Europe received no further correspondence from the elusive Prester John.

But the legend of Prester John did not die. Europeans at first imagined his kingdom to be somewhere in India. Then, when India began to be a little better known, they started placing Prester John's kingdom in the uncharted expanses of central Asia. In the thirteenth century, the Mongol Empire rose to power across Asia. For the first time, European ambassadors were able to travel unhindered to the Mongol homeland in the distant East, and they looked for Prester John as they went. The first of these emissaries was an Italian monk

The Mongol warlord Genghis Khan (at center, holding a scepter) created an empire that covered most of Asia. Accounts of his power probably contributed to the Prester John myth.

named John of Plano Carpini, who crossed Asia in 1245-46. He wrote an account of his travels in which he admitted that he had not actually *seen* Prester John's kingdom, but he believed it lay somewhere just off his route.

Two other monks, William of Rubruck and Odoric of Pordenone, were less certain. They said that *if* Prester John existed, the stories told about him were greatly exaggerated. Marco Polo, the best-known and most wide-ranging of the medieval European travelers, spent the years from 1275 to 1292 in China. When he returned to his native Venice, he declared that there *had* been a Prester John, but that he had died in battle some years earlier. It is now known that Polo confused Prester John with Ong Khan, a Mongol warlord of the early thirteenth century who was neither rich, nor particularly powerful, nor a Christian.

These early European travelers across Asia failed to bring back accurate reports of Prester John's kingdom, but they did gather a wealth of information about the peoples, trade routes, cities, and products of central Asia and China. This information smoothed the way for many later travelers. It also inspired countless journeys and voyages of exploration eastward. In particular, Marco Polo's experiences in China, recounted in a volume called *Description of the World* (sometimes also known as the *Travels*), fed Europeans' growing appetite for news of the mysterious East. His description of his life in the court of the Great Khan of the Mongols is a priceless account of the inner workings of the largest land empire the world has ever known.

Still, Prester John haunted the European imagination. By the end of the fourteenth century, European geographers were fairly certain that Asia did not contain John's kingdom, but they could not bear to give up the long-cherished legend. They decided that Prester John's kingdom must lie somewhere in the remote interior of another huge, almost entirely unknown continent—Africa.

In 1459 a Venetian monk named Fra Mauro produced a map that placed Prester John's kingdom in the East African realm of Abyssinia (today called Ethiopia), which was indeed ruled by Christian princes from the fourth century until the twentieth. In the part of his map that shows Abyssinia, Fra Mauro wrote, "Here Prester John makes his principal residence."

The rulers of Portugal believed that Prester John and the Abyssinian emperor—about whom little was known other than vague rumors—must be one and the same. Starting in the early fifteenth century, the Portuguese launched a series of explorations south along the west coast of Africa. Their main goal was to find a sea route around Africa to the Indies, but they also hoped to make contact with Prester John somewhere in the interior of Africa.

In 1483 a Portuguese navigator named Diogo Cão, venturing farther south in the Atlantic than any European had gone before, came upon the mouth of the Congo River (today sometimes called the Zaire). He learned from the local Africans that there was a large state called the Kingdom of Kongo some distance upstream. Its capital was a majestic city upon a mountain, and its ruler was a powerful king called the Manikongo. Cão jubilantly decided that he had found Prester John at last.

King João II of Portugal wanted Prester John as an ally. João's determination to find the elusive Prester John led Portuguese explorers into the Congo River basin and the mountains of what is now Ethiopia.

Cão desperately wanted to contact Prester John, but he was afraid to send any of his Portuguese crewmen into the unknown and forbidding jungle. Instead, he sent four of the black slaves from the Guinea coast who had come on the voyage; they spoke Portuguese as well as their native languages and had been baptized as Christians.

He equipped them with trade goods as gifts for Prester John and sent them off upriver, promising to pick them up later. Then Cão set off southward once again to continue his exploration of the African coastline, as ordered by King João II of Portugal.

A few weeks later Cão stopped at the Congo mouth on his way home to pick up the four "ambassadors" he had sent upriver. To his fury and dismay, the four were nowhere to be seen, and Cão decided that the Africans must treacherously have captured or killed them. He seized four Africans, in revenge or as hostages for his slaves' return, and sailed to Lisbon.

Cão had done the Manikongo an injustice. The four slaves were greeted courteously and treated well in the Kingdom of Kongo. Perhaps more surprising was the treatment the four African natives received in Lisbon. The Portuguese had been slave trading along the Guinea coast for years, and they had come to regard all Africans as slaves. But King João decided to treat Cão's captives with great kindness, as though they were ambassadors and not prisoners, so that when he allowed them to go home they would make a good report to their master, the Manikongo.

The four Africans were therefore given apartments in the royal palace, dressed in European clothing made of the finest silks and velvets, treated to elaborate ceremonial feasts, and taken on tours of Portugal. They were also tutored in Portuguese and in Christian dogma, and they were given lavish gifts to take back to the Manikongo.

King João was certain that either the Manikongo was Prester John, or at least that the Manikongo knew the elusive Christian king and could help the Portuguese find him. João and his geographical advisers were also convinced that the Congo River flowed into the Atlantic Ocean from the highlands of Abyssinia on the other side of Africa, and that it would therefore provide a highway straight to Prester John's kingdom. They were quite wrong, for the Congo originates in central Africa, far south of Abyssinia, and flows north and west in a great semicircle through the forested heart of the continent before reaching the sea. Nevertheless, King João optimistically sent Cão out again in 1485, to extend his exploration of the African coast and to carry the African visitors home.

Cão returned the four Africans to the very place he had snatched them from, where they were joyfully greeted by their friends and

(continued on page 57)

MAN'S CHANGING VIEW OF THE WORLD

Exploration and mapmaking went hand in hand. Early geographers collected travelers' tales and began to piece together a picture of the world. Many explorers were inspired by ancient maps; Christopher Columbus, who made history's biggest serendipitous discovery, carefully studied every world map he could find before setting sail. And the explorers, in turn, brought back information that was used to make new and better maps. Over the centuries, mapmakers patiently drew and redrew the world, recording each new discovery and reflecting our growing knowledge of the world.

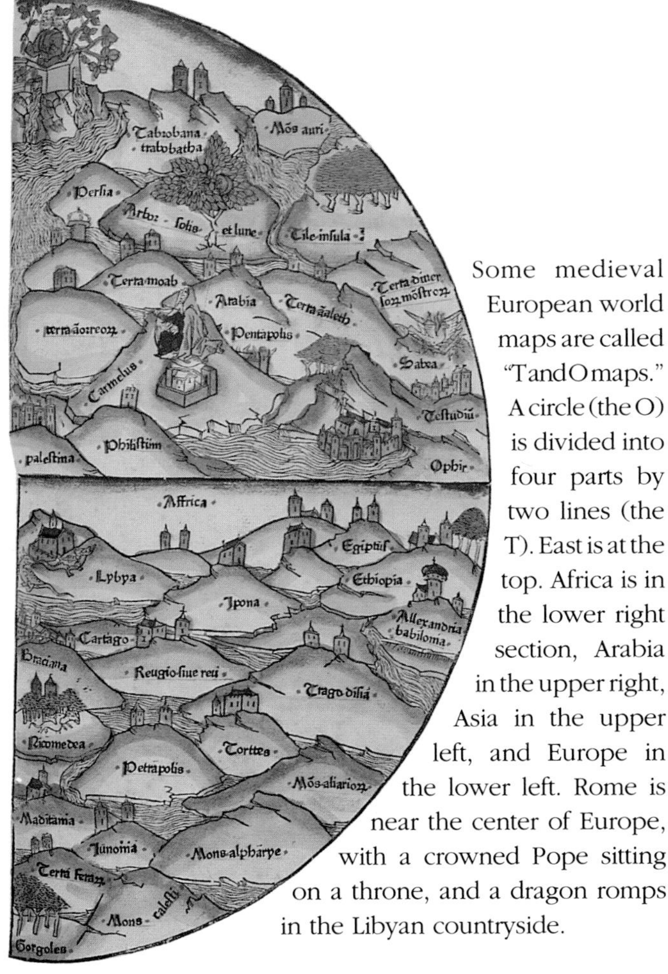

Some medieval European world maps are called "T and O maps." A circle (the O) is divided into four parts by two lines (the T). East is at the top. Africa is in the lower right section, Arabia in the upper right, Asia in the upper left, and Europe in the lower left. Rome is near the center of Europe, with a crowned Pope sitting on a throne, and a dragon romps in the Libyan countryside.

CAVRVS·CHORVS·VEL·IAPIX·SIVE·ARGESTES
CIRCTVS
VEL·TRESTIAS
FAVONIVS·ZEPHIRVS
mare glaciale
EVROPA
Pontus euxinus
ARIA
LIBIA INTERIOR
AFFRICA
ARABIA FELIX
Gradus equinoctialis
ETHIOPIA INTERIOR
Terra incognita secundum ptholomeum
Sinus Barbaricus
Tropicus
Gradus
AFRICVS·VEL·LIBS
LIBONOTVS·EVROAVSTER
AVSTER·VEL·NOTVS

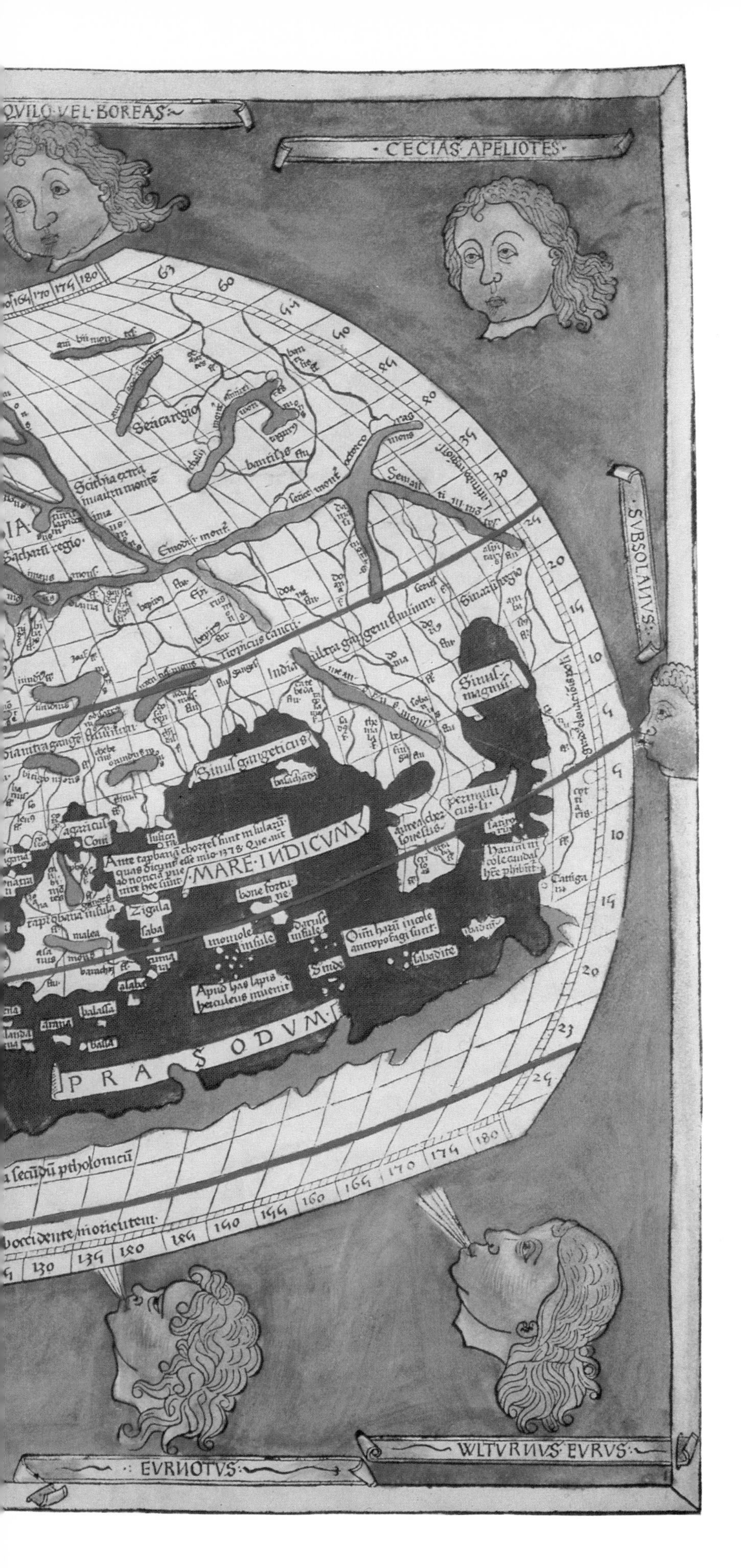

Claudius Ptolemaeus, known as Ptolemy, was a Greek geographer of the second century A.D. who greatly influenced later mapmakers. This Ptolemaic world map, ornamented with the twelve wind gods of ancient mythology, was published in 1482. The African coast curves east to meet the coast of Asia, turning the Indian Ocean into a giant lake. Soon explorers would sail around Africa, disproving Ptolemy's geography.

Found in a Paris shop in the nineteenth century, this is the first world map to show the Americas. It was drawn on oxhide in 1500 by Juan de la Cosa, who sailed with Columbus. The Americas appear as a scattering of islands and coastal points; the blue-green mass in the west is unexplored territory. The ships sailing around the bottom of Africa represent the fleet of the Portugese navigator Vasco da Gama.

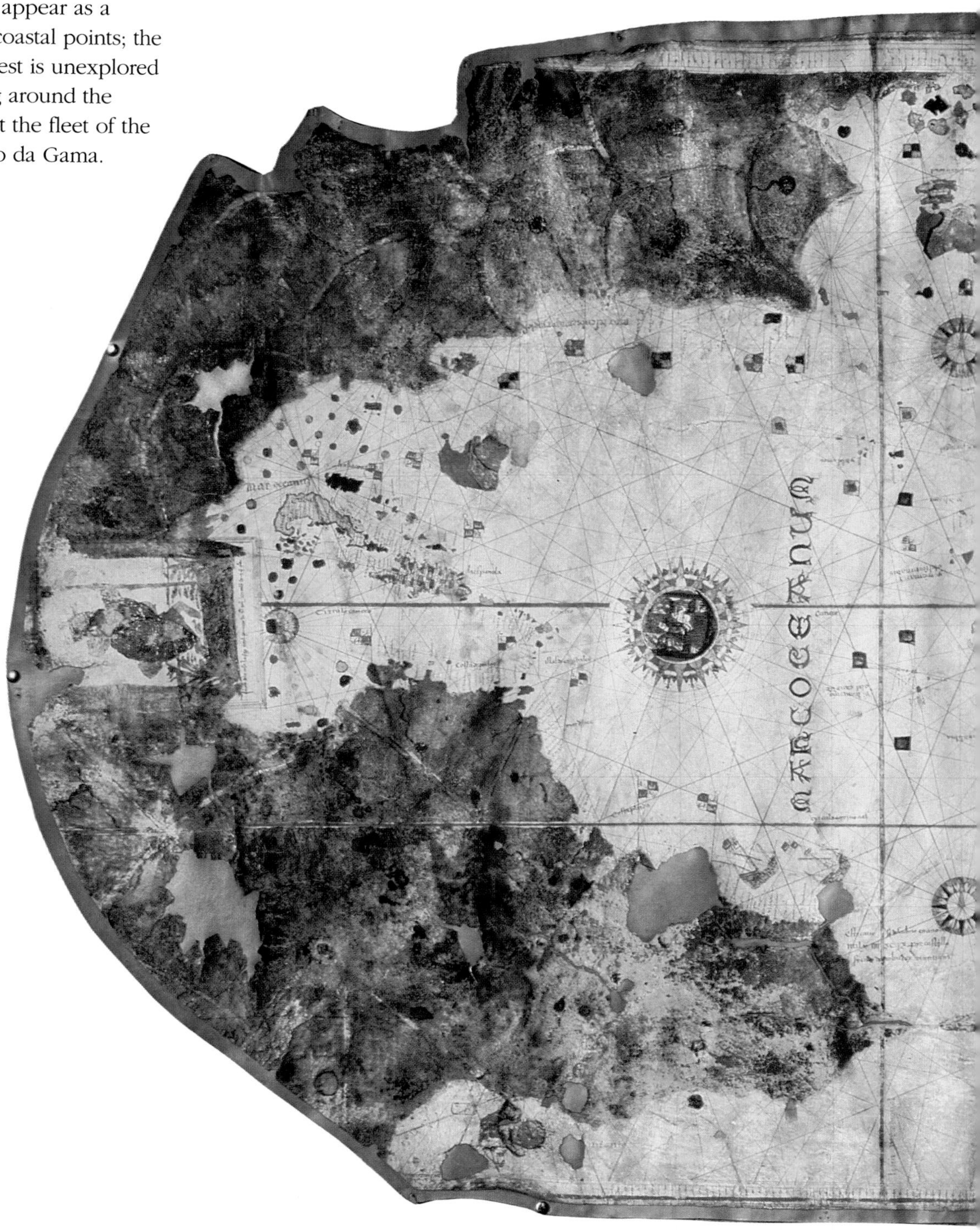

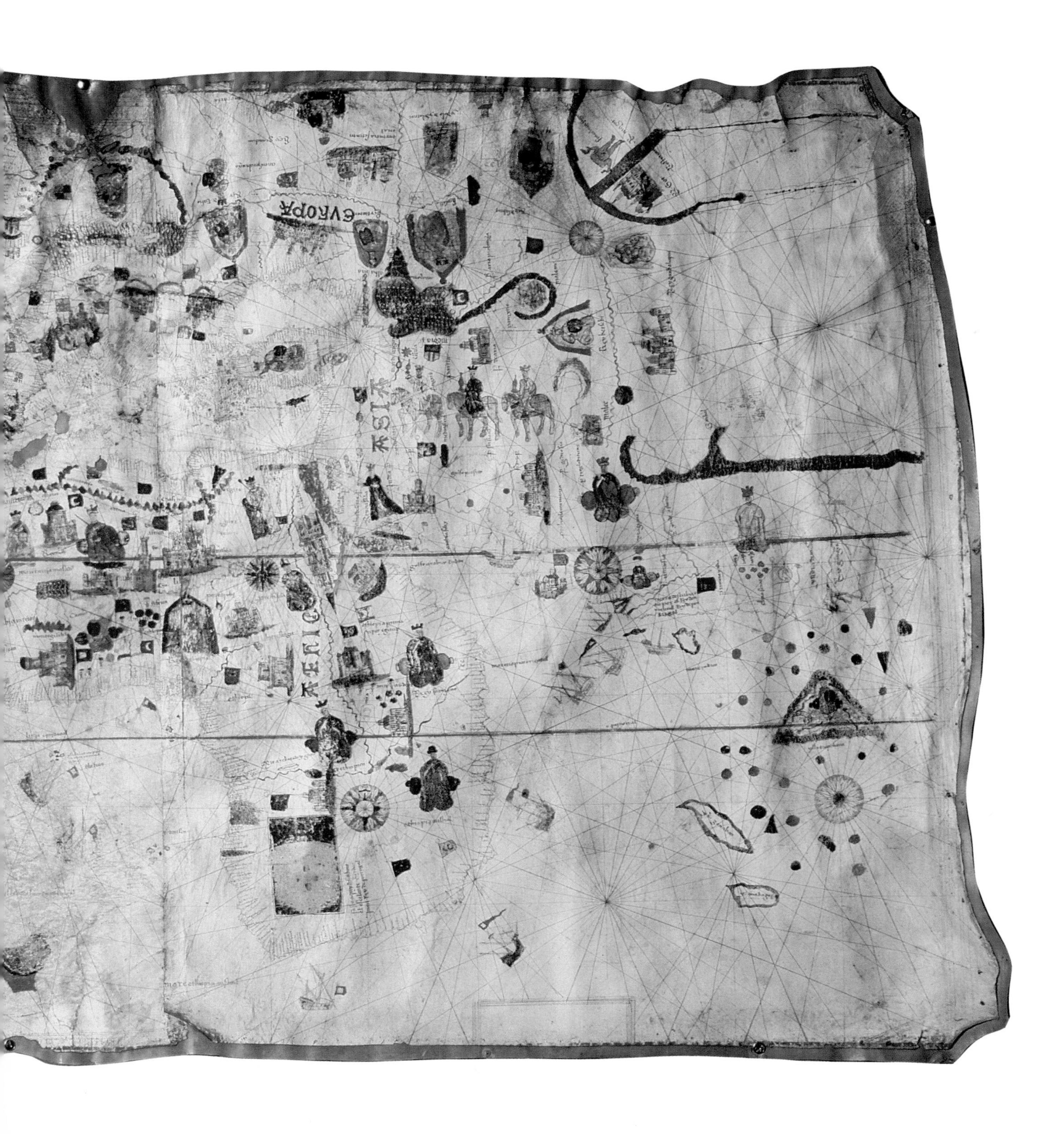

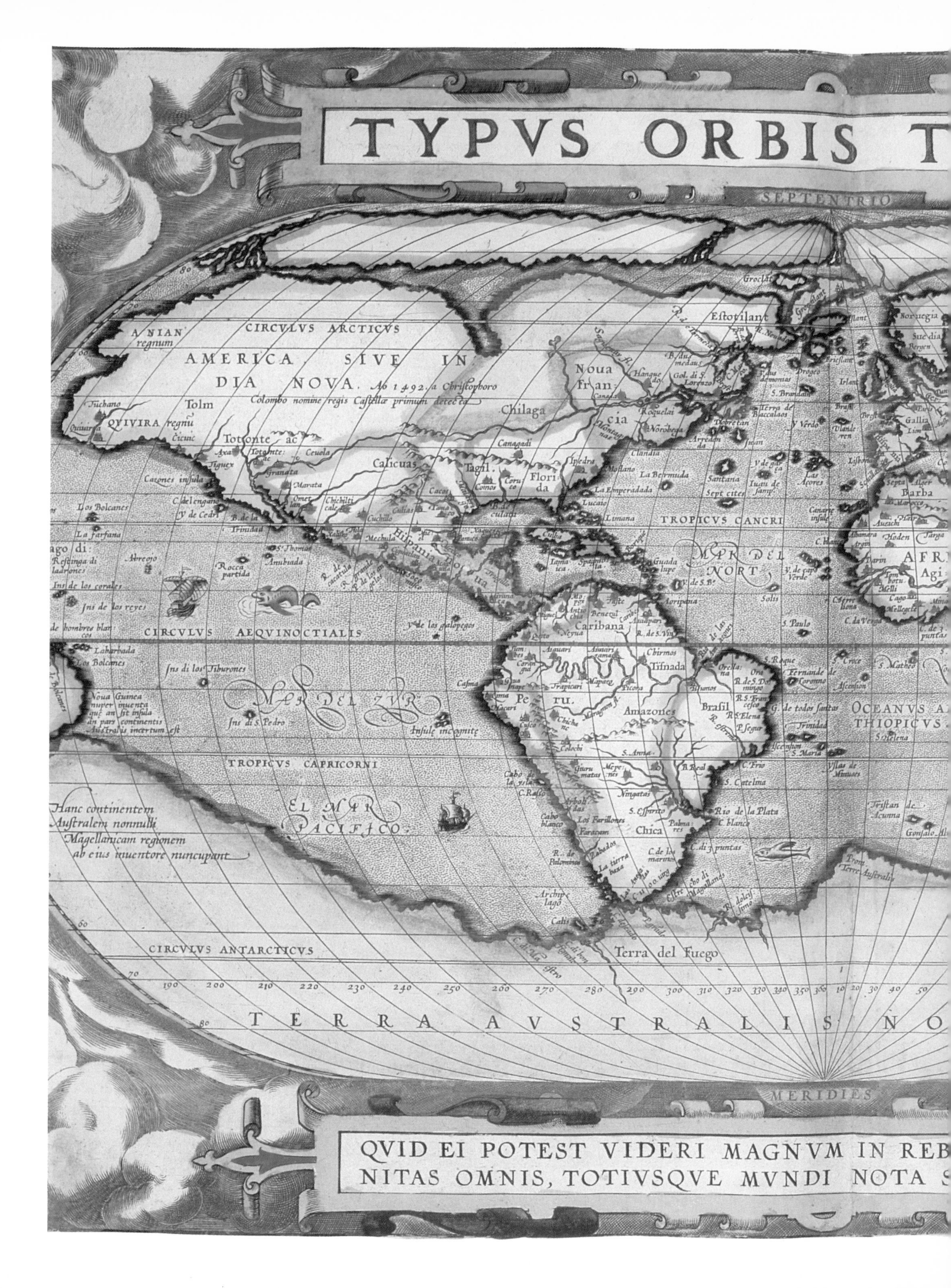

TYPVS ORBIS T
SEPTENTRIO
CIRCVLVS ARCTICVS
AMERICA SIVE INDIA NOVA.
ANIAN regnum
QVIVIRA regnu
Tolm
Chilaga
Noua Francia
Estotilant
Groclat
Canagadi
Calicuas
Tagil
Florida
Hispania noua
La Bermuda
TROPICVS CANCRI
MAR DEL NORT
Los Bolcanes
CIRCVLVS AEQVINOCTIALIS
Caribana
Peru
Amazones
Brasil
MAR DEL ZVR
OCEANVS AETHIOPICVS
Insulae incognitae
TROPICVS CAPRICORNI
EL MAR PACIFICO
Chica
Rio de la Plata
Hanc continentem Australem nonnulli Magellanicam regionem ab eius inuentore nuncupant
Noua Guinea nuper inuenta quę an sit insula an pars continentis Australis incertum est
Terra del Fuego
CIRCVLVS ANTARCTICVS
TERRA AVSTRALIS NO
MERIDIES
Barba
AFR
QVID EI POTEST VIDERI MAGNVM IN REB
NITAS OMNIS, TOTIVSQVE MVNDI NOTA S

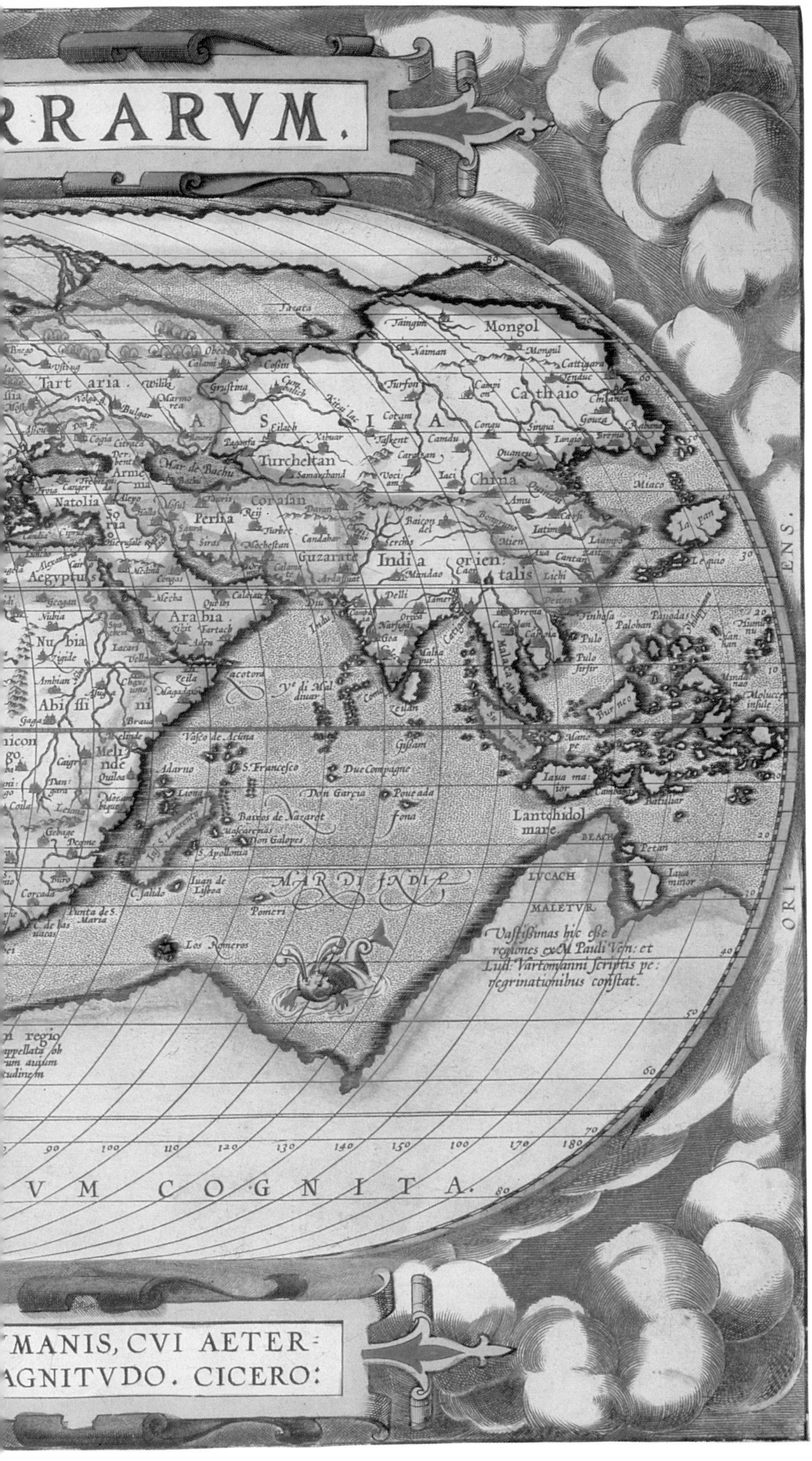

Abraham Ortelius, a sixteenth-century map publisher, produced the first modern atlas, or book-length set of maps. The 1586 edition of the atlas included this world map. Ortelius agreed with Ptolemy that a huge continent lay waiting to be discovered in the southern half of the globe. He labeled it Terra Australis Nondum Cognita—Latin for "the southern land not yet known."

The ancient Greeks are thought to have made globes, but this is the oldest known surviving globe. Made in 1492 by a German traveling merchant and amateur geographer named Martin Behaim, it was called Behaim's *Erdapfel*, or earth apple. Adorned with elephants, castles, and mermaids, Behaim's globe reminds us that well before Columbus sailed, people knew that the earth is round—like an apple.

(continued from page 48)

families. They then set off into the interior, promising to bring Cão's four slaves back from the Manikongo. While waiting for their return, Cão and his men sailed a distance up the Congo River—scouting the way, they thought, for the flotilla that would surely be sent to Prester John's kingdom. But after about 100 miles (160 kilometers) they found that the river was blocked by a roaring mass of waterfalls and rapids. Later explorers called these cataracts the Cauldron of Hell. No ship could sail beyond this point. Cão turned back to the coast.

Soon Cão's four slaves returned from Mbanza Congo, as the Manikongo's capital city was called, and with them came a number of envoys from the Kingdom of Kongo to the Kingdom of Portugal. They were the sons of Kongo nobility; one of them was a prince named Nsaku. Cão took them back to Portugal, where they were treated royally. Nsaku became a Christian, and King João attended the ceremony of his baptism, at which he took the name Dom João

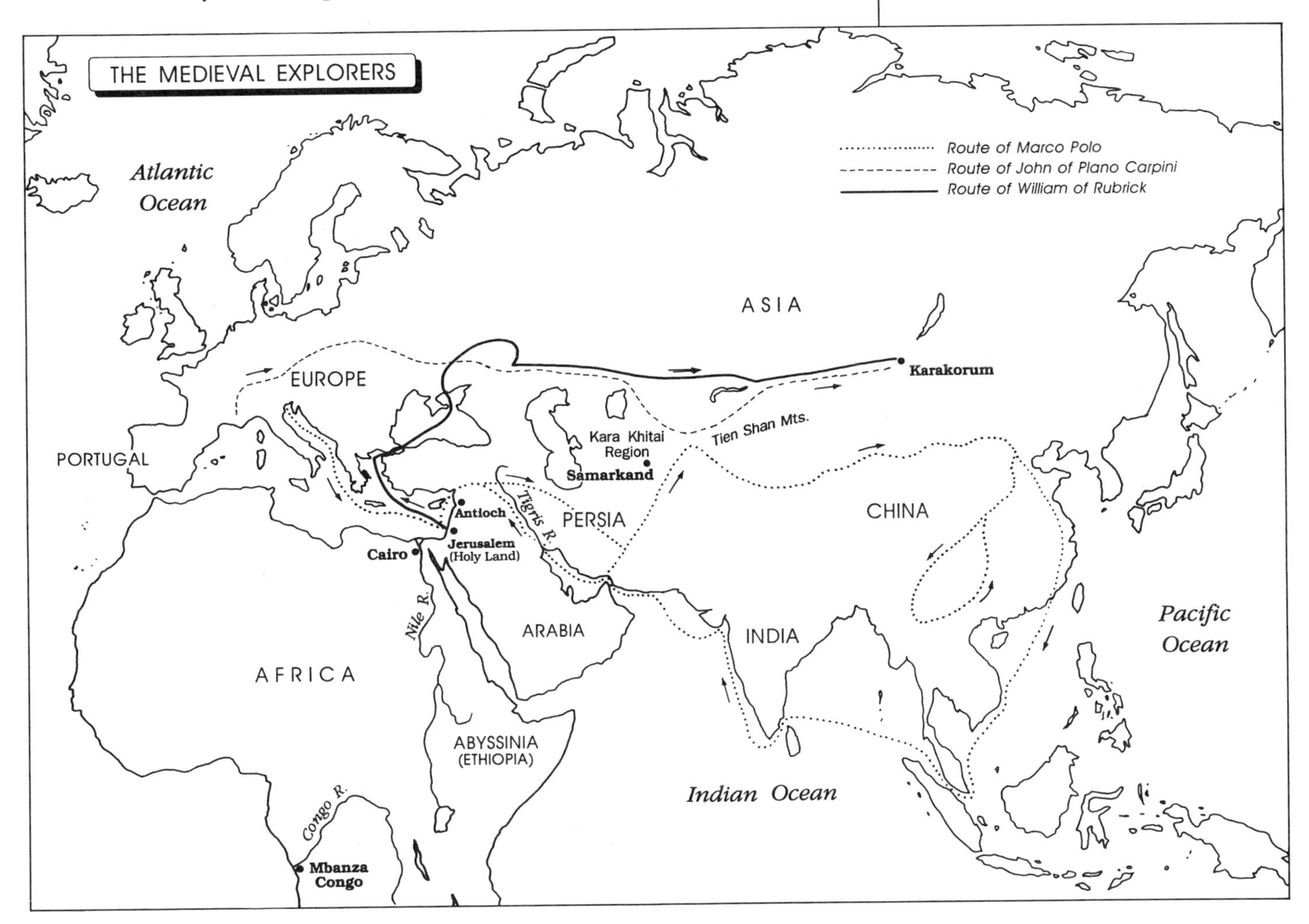

da Silva (Sir John of the Woods). Tragically, he died of the plague within a few years.

King João sent an expedition to Mbanza Congo. In 1491, after an arduous journey through dense forest and over rugged mountains, the Portuguese reached the capital city, which was located about where the city of São Salvador do Congo now stands in the country of Angola. The city and its ruler disappointed the Portuguese. Although they admitted that Kongo had a very impressive and highly developed culture, the Europeans had been expecting the extravagant riches and marvels that were traditionally associated with the Prester John story. But they found no treasures, and no Christians, along the Congo River. At once they set about building churches, baptizing the Africans, and destroying sacred statues and other objects of native worship; they also sent out several expeditions in search of Prester John, but none of these searchers ever returned.

The Portuguese, with their greedy, blustering, and intolerant ways, soon wore out their welcome in Kongo. In 1495 the Manikongo renounced Christianity and tried to throw off the yoke of Portuguese domination. But the people of Kongo and the neighboring African peoples found that the Europeans, once welcomed, were difficult to get rid of. A large part of Africa around the Congo River remained under Portuguese control for centuries. Angola, where once the Manikongo ruled, was a Portuguese colony until 1975.

The Congo River, although it was a source of slaves, ivory, and other goods prized by the Europeans, was a dead end as far as reaching Prester John in faraway Abyssinia was concerned. Finally, however, someone did reach "Prester John's kingdom."

In 1487 King João II of Portugal sent one of his most experienced spies, a man named Pero da Covilhã, who spoke fluent Arabic, on a mission of industrial espionage. Covilhã's orders were to make his way to India and find out as much as he could about the spice trade.

Covilhã disguised himself as a traveling Muslim trader with a cargo of honey. From Cairo, Egypt, he managed to make his way undetected all the way to India on an Arab merchant ship. He snooped around the spice ports of India, then visited the east coast of Africa, returning to Cairo in 1489 or 1490. He was ready to return to his wife and children in Portugal, but King João had other plans. In Cairo, Covilhã received new orders. He was to go to Abyssinia and make a treaty of friendship with Prester John.

No details of Covilhã's journey to Abyssinia survive, but it is certain to have been a remarkably difficult one, through territory unknown to Europeans at the time. He reached the Abyssinian court in 1493. He was treated well, but the emperor of Abyssinia would not let him leave. When ambassadors from Portugal came to Abyssinia thirty years later, they found Covilhã still there, an honored prisoner.

Although the Prester John of legend managed to elude everyone who sought him, he continued to occupy a prominent place in the European imagination, and journeyers returning from distant places were often questioned about him. Travelers sometimes like to dazzle the stay-at-homes with tales of amazing sights and exciting exploits, and one such traveler impressed quite a few people with his story of a visit to Prester John's kingdom. The traveler was Niccolò de' Conti, a fifteenth-century Venetian who spent twenty-five years roaming Asia for pleasure and profit.

Conti spent part of his youth in Damascus, a Syrian center of the spice trade, and there he learned Arabic. Leaving Italy in 1414, he made his way to Baghdad, in present-day Iraq, and then east into Persia. He adopted the Persian language and style of clothing, and he went into partnership with some Persian merchants. This partnership sent him into India on business.

Conti lived in India on and off for several decades, but during this time he also visited Ceylon, Sumatra, Burma (now called Myanmar), Java, and Southeast Asia. He married an Indian woman and together they had four children. In the 1440s, he went west to Arabia. Then he moved on to Cairo and finally returned to his native Venice in 1444.

On the way to Venice from Cairo, Conti stopped at Jerusalem. There he met a Spanish traveler who was proud of having come all the way from Spain to Jerusalem, across the length of the Mediterranean Sea. Hearing this, Conti could not refrain from boasting about his own travels. He described the cinnamon forests of Ceylon, the gold and the cannibals of Sumatra, and the temples of Java. He told of the elephants, rhinoceroses, and tattooed people he had seen in Burma. He recounted his search for jewels along India's Malabar Coast. And then he spun a long and elaborate fantasy about his visit to the one place that was guaranteed to excite any listener—Prester John's kingdom. Conti's Spanish acquaintance solemnly wrote the whole thing down, fables as well as facts.

(previous pages) Prester John was not the only geographical phantom of the Middle Ages. Europeans believed that unexplored lands were inhabited by all kinds of strange beings, including people with the heads of dogs and headless people whose faces were in their chests.

Another, more sober, version of Conti's story appeared later. He was in disfavor with the Roman Catholic Church for having renounced, or given up, the Christian faith during his sojourn in India. In the eyes of the church, this made him a heretic, subject to serious punishment. Pope Eugenius IV allowed him to redeem himself by narrating an account of his adventures to a papal secretary. This time Conti provided a more accurate version of his history, and the result was a document full of sound information about the geography and people of fifteenth-century India and points beyond.

The most important legacy of Conti's travels is not his frivolous tale of Prester John. It is the speculation he offered that Europeans could reach India by sailing south and east around Africa. This notion contradicted Ptolemy's *Geography,* but it was nevertheless incorporated into several fifteenth-century maps and discussed fairly widely by geographers and mariners. It helped to launch the great voyages of exploration that set sail from Portugal and Spain in the late fifteenth and early sixteenth centuries.

The myth of Prester John had remarkable longevity. In 1573 the Dutch mapmaker Abraham Ortelius labeled Abyssinia "the empire of Prester John." And stray references to the mysterious monarch kept cropping up on maps and in travel chronicles as late as the early eighteenth century.

But who *was* Prester John? Or, more significantly, how and why did the legends about him arise?

Undoubtedly one source of the Prester John myth was the Christian communities that really did exist throughout Asia from early medieval times. These communities consisted of Armenian Christians (originally from what is now Turkey) and Nestorian Christians (originally from what is now Syria) who had split away from the Catholic Church over differences in dogma. Persecuted as heretics, the Armenians and Nestorians fled into remote places in central Asia and China. There were also a few Christians in India. Rumors of these small, isolated Christian communities drifted back to Europe in later centuries and helped create the myth of a Christian kingdom in Asia.

As for Bishop Otto's account of Prester John and the battle against the Saracens, he may have been referring to an Asian warlord named Yelui Tashi, a chieftain of the Kara Khitai people. In the twelfth century the Kara Khitai lived north of the Tien Shan Mountains, above present-day Pakistan. Yelui Tashi was probably a Buddhist, but it is just possible that he was a Nestorian Christian. In 1141 he defeated a Muslim army in a great battle at the ancient central Asian city of Samarkand. This may be the battle that was described to Otto by his friend the Syrian bishop.

The letter of 1165 was a forgery. Despite the earnest efforts of scholars over the centuries, no one knows who wrote it, or why, or even what language it was written in. The three original letters were lost long ago, and only copies remain.

Perhaps the letter was a prank, a colossal leg-pull. Perhaps it was meant as entertainment, the twelfth-century equivalent of an adventure movie. In his 1954 book *Conquest by Man,* the German historian Paul Herrmann advanced another theory. He suggested that the letter was a political treatise. Passages in the letter say that Prester John's kingdom knew no crime, no poverty, and no private property (except for John's, of course). Herrmann thought that the letter's anonymous author may have used Prester John as a gimmick to draw attention to his description of a well-governed, perfect state, exposing the injustices and inequalities of the European states by contrast.

The identity and motives of the letter writer will probably remain a mystery forever. But the Prester John letter lit a flame of hope and belief that burned for centuries and stimulated many journeys and voyages of discovery. Great stretches of Asia and Africa were explored in the long quest for Prester John's realm. "We will never know," says historian Daniel J. Boorstin in *The Discoverers,* "how many true believers were seduced to search for the mythic kingdom."

CHAPTER 6

El Dorado and the Fountain of Youth

According to legend, every morning servants anointed El Dorado with sticky resin and sprinkled him with gold dust.

The legend of Prester John belonged to the Old World, but the Spanish conquerors of the New World were soon chasing a whole new set of rumors and myths. Two of the most enduring of these were El Dorado and the Fountain of Youth.

After they had overcome the Aztec and Inca empires, the Spanish looked around for new conquests. The Aztecs and Incas had possessed gold and silver almost beyond imagining, and the Spanish were very ready to believe that other, perhaps richer, kingdoms lay hidden in the unexplored regions of the Americas.

The Spanish conquistadors in South America began to pick up rumors from the Indians of a king they called El Dorado, "the Golden Man." El Dorado, it was said, ruled over a tribe that worshiped the sun. Every morning the king covered himself with gold dust until he shone like the sun, then rinsed the dust off in a sacred lake. His people also threw gold artifacts into the lake.

A lake full of gold dust and jewelry sounded good to the Spanish, and the story grew more fabulous each time it was told. Before long the storytellers had transformed El Dorado from a golden man into a golden city, or even a kingdom with *two* cities made of solid gold. Everyone in El Dorado wore gold dust every day, the streets were paved with gold, and the doorstops were diamonds.

The golden city of El Dorado lies conveniently close to a harbor in this imaginative rendering (opposite). But explorers who searched for the city found only fever-ridden swamps and jungles.

Many conquistadors went looking for El Dorado, which was said to lie somewhere in the all-but-inaccessible tangle of mountains and jungles where Colombia, Venezuela, and Brazil come together. When Gonzalo Pizarro set off on his expedition across the Andes Mountains—the same expedition from which Orellana launched his unexpected river journey—he was hoping to find El Dorado. Nor were the Spanish the only ones to seek the source of the legend. Sir Walter Raleigh, the British adventurer best known for founding the short-lived colony on Roanoke Island, North Carolina, traveled up Venezuela's Orinoco River in 1595 and again in 1617 in search of El Dorado.

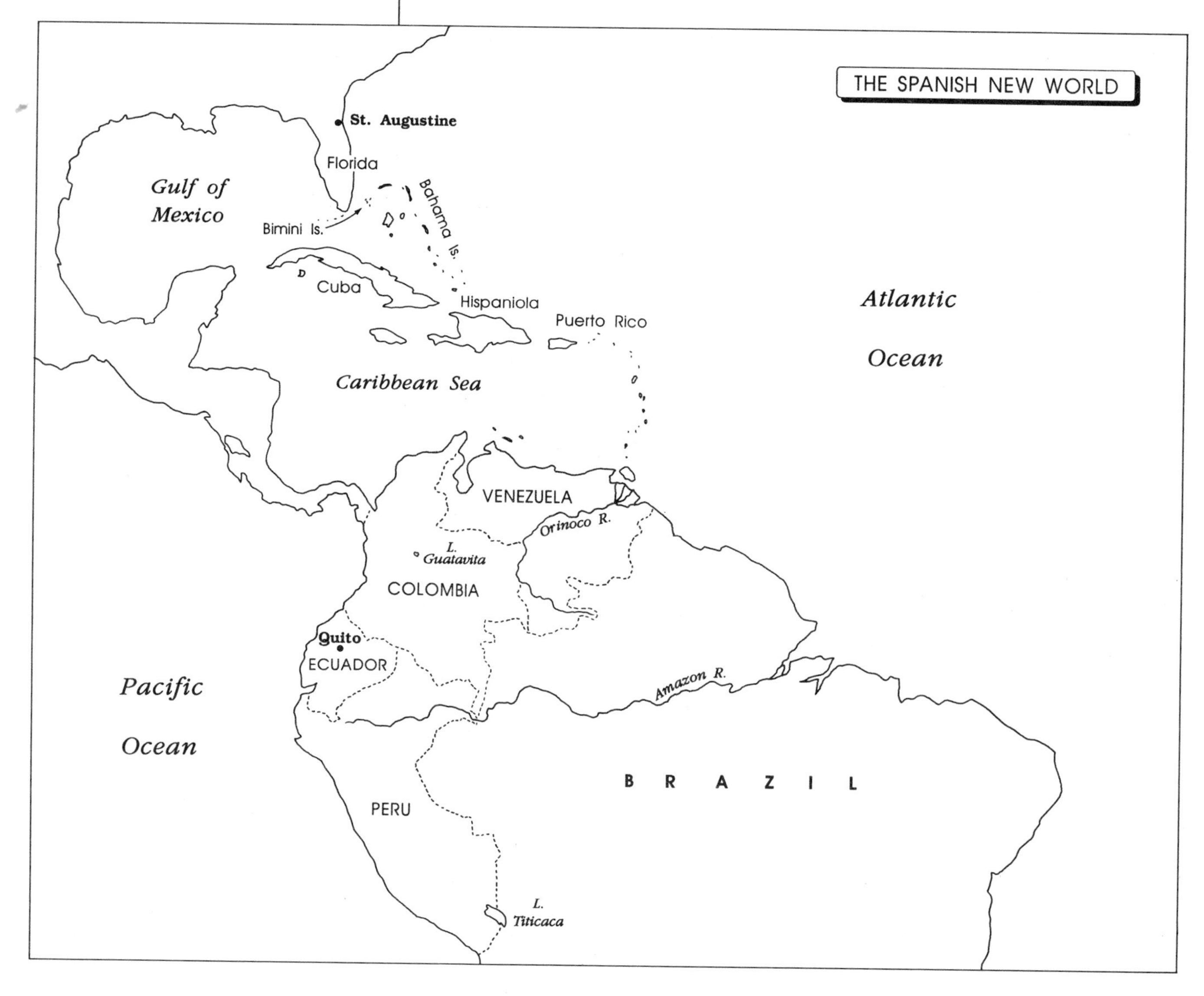

The British adventurer Sir Walter Raleigh looked for El Dorado along Venezuela's Orinoco River. His men battled alligators, heat, and starvation, but found no gold.

Over the next several centuries, many expeditions disappeared into the pathless rainforest or the high plateaus of South America, all seeking the golden city of El Dorado. None of them found it, which is not surprising; it does not exist.

But there *is* a basis for the original El Dorado legend. Lake Guatavita, near the capital city of Bogotá in present-day Colombia, was part of the territory of the Muisca Indians, sometimes called the Chibcha. The Muisca possessed gold, and they were known to make golden offerings to their gods. By the mid-sixteenth century Guatavita had become identified with the lake of the legend. In 1562 a Spanish conquistador named Antonio de Sepulveda made the first attempt to drain the lake, and he found a small but significant amount of gold.

Over the years, many further efforts have been made to get at the gold that is thought by some to lie buried in the silt on the bottom of Lake Guatavita. Attempts have been made by Spanish adventurers, British engineering companies, and American scuba divers. Some golden artifacts have been recovered, but fortune hunting is now prohibited by the Colombian government, which protects the lake as an archaeological site.

Unimaginable wealth is a delightful dream, and so is prolonged youth in which to enjoy that wealth. The Spanish pursued both dreams in the New World. Although the dream of El Dorado never came true for the Spanish, they *did* reap immense riches from the Aztecs and the Incas. They had no luck at all, however, with the Fountain of Youth.

Juan Ponce de León was a Spanish conquistador who sailed on Columbus's second voyage (1493-96). During that voyage, on his way to set up the colony on Hispaniola, Columbus anchored off the coast of the island that is now called Puerto Rico and claimed the land for Spain. His fleet spent two days there before moving on to Hispaniola.

Ponce de León returned to the island fifteen years later, in 1508, determined to conquer it and turn it into a prosperous Spanish colony. He gave the island its name, which means "Rich Port" in Spanish, and he rapidly brought its Native American people, the Arawak, under control. As a reward, the Spanish crown named him governor of Puerto Rico.

From the Arawak, Ponce de León heard legends about a place called Bimini, where there was said to be a fountain whose waters had the power to cure illnesses. Anyone who drank from the fountain, the Indians said, remained healthy and youthful. These tales filled

The mythical Fountain of Youth was sometimes believed to be the source of the four rivers of Paradise mentioned in the Bible. Ponce de León found neither Paradise nor eternal youth—only mosquitoes, mangrove swamps, and death.

the conquistador with eagerness to find the fountain and taste its waters. Bimini was rumored to lie north of Puerto Rico, and in March 1513 Ponce de León set off in that direction with three ships.

On May 27 he sighted land, which he named Florida because the day was Easter Sunday, called Pascua Florida in Spanish. He put ashore a few days later on the eastern coast of the Florida peninsula, which he believed to be an island, and claimed the place for Spain. He went on to explore both the Atlantic Ocean and the Gulf of Mexico coasts of Florida, but he found no Fountain of Youth, only dense mangrove swamps.

Ponce de León did not give up on Florida or the Fountain of Youth. He returned to Spain in 1514 and received permission from the crown to start a colony in what was called the Island of Florida. In 1521, with high hopes of establishing a colony and tracking the rumors about the elusive fountain to their source, he returned to Florida, where he was wounded in a fight with the local Native Americans. The injury was serious, and there was no water from a miraculous fountain to cure it. Ponce de León made it as far as Cuba, and there he died.

Nothing more was heard of the Fountain of Youth. As for the legendary land of Bimini, the name was given to two tiny islands in the northern part of the Bahama Islands chain. And as for Florida, it remained a Spanish possession for many years. The city of St. Augustine rose on the spot where Ponce de León first landed, and it is the oldest continuously occupied European settlement in North America.

Florida became the property of the United States in 1819, and in the twentieth century it has become one of the country's foremost vacation spots and retirement havens. College students flock to its beaches during their spring vacations, and hundreds of thousands of older Americans spend their retirement years basking in its sunshine. It is not quite the Fountain of Youth that Ponce de León imagined, but perhaps he would approve nonetheless.

Ponce de León came to the Caribbean on Columbus's second voyage. Years later he conquered Puerto Rico and began his search for the Fountain of Youth.

CHAPTER 7

The Seven Cities of Cíbola

The conquistadors were impressed by the American bison, which roamed the West in vast herds.

South America had its rumored city of gold, but North America did even better. It had the legend of the Seven Cities of Cíbola, seven mighty cities whose walls were said to be made of gold and studded with turquoise. The kingdom of Cíbola was supposedly located somewhere north of the Spanish colony in Mexico, in the region that is now the southwestern United States.

The Cíbola legend was started by Friar Marcos of Nice, a monk with an overactive imagination. In 1539 he traveled from Mexico north into present-day Arizona and New Mexico—territory that was completely new to Europeans at the time. He brought back the story of the Seven Cities of Cíbola, although it is not clear whether he actually claimed to have seen the cities himself or merely to have heard about them. He reported that the Indians of Cíbola used gold for everything, even to make "little blades with which they wipe away their sweat."

Luxuriating in the wealth wrested from the Aztecs, the Spanish viceroy in Mexico listened with favor to Friar Marcos's tale and promptly assigned a provincial governor named Francisco Vásquez de Coronado to lead an expedition to Cíbola. Coronado had no trouble signing up 336 soldiers who, according to one account, "were in Mexico, and, like a cork floating on water, went around with nothing to do." The expedition also included 1,300 Indians, 5 priests, and 559 horses. Friar Marcos was their guide. They set out northward in February 1540.

Before long the soldiers began to grumble at Friar Marcos, because the hot desert of northern Mexico was nothing like the lush plains

Coronado's expedition to Cíbola, as envisioned by the nineteenth-century American artist Frederic Remington (opposite).

THE CORONADO EXPEDITION

he had described to them. The grumbles turned to cries of rage and disappointment when they reached the first Indian settlement. Instead of the populous city Marcos had promised, it was an abandoned ruin. One soldier remarked, "The curses that some hurled at Friar Marcos were such that God forbid they may befall him." The monk told them that Cíbola was farther north. They kept going.

After four and a half months they had covered 1,500 miles (2,400 kilometers) and were beginning to fear starvation. Then they came to what Friar Marcos identified as the first of the Seven Cities of Cíbola. It was an adobe pueblo dwelling called Hawikuh, a Native American settlement on what is now the Zuni Indian Reservation in western New Mexico. There was no gold; there were no turquoise-studded walls. The fabled city, said one member of the party, was "a little cramped village looking as if it had been all crumpled up together."

The Spanish were unable to appreciate the ingenuity of the pueblo dwellings. They were angry when they learned that the other six cities were much like Hawikuh. Friar Marcos was in some danger from the wrath of the soldiers, so Coronado sent him back to Mexico. He also sent a letter to the viceroy, in which he said bitterly that Marcos had "not told the truth in a single thing that he said."

After some initial fighting with the Zuni, in which Coronado was wounded, the Spanish commander and his men camped in the pueblo; eventually they established friendlier relations with the Zuni. Coronado sent parties of men out to explore in several directions. Then he led his men on a long, winding journey to the plains of eastern New Mexico, where they saw immense wild herds of huge, shaggy cattle. "There are such quantities of them that I do not know what to compare them with unless it be the fish of the sea," one Spaniard later recalled. Coronado's men had encountered the vast bison herds of the American West.

The expedition spent the winter in a pueblo and set off the next spring on the heels of yet another golden rumor. An Indian nicknamed the Turk had told Coronado of a kingdom called Quivira, located in present-day Kansas. In Quivira, the Turk claimed, everyone was rich enough to eat from golden plates. That glittering image drew Coronado across the dusty plains of Texas and into the heart of Kansas. The Spanish approved of the fertile, well-watered Kansas prairie, but they were grievously disappointed in Quivira, which turned out to be a humble settlement of Wichita Indians. Once again, no gold. The Turk admitted that he had lied about the gold, and Coronado had him strangled. Then he began the long march home.

The Spanish expedition spent the winter of 1541-42 in the pueblo again. Coronado returned to Mexico in 1542, in disgrace because he brought with him "neither gold nor silver nor any trace of either." But the Coronado expedition *did* discover one glorious treasure, although Coronado himself did not see it, and it was not the sort of treasure that could be carried back to Mexico in a saddlebag.

One of the scouting parties that Coronado sent out from Cíbola in the summer of 1540 reached a high plateau about 200 miles (320 kilometers) to the northwest. Riding through the pine trees of this plateau, López de Cárdenas and his twenty-five companions came to a sheer cliff and found themselves, to their astonishment, gazing out across a wide, deep canyon whose walls were brilliantly streaked as if by red and yellow paint. The opposite side of the canyon seemed to float in the air miles away, and its floor, where a tiny streak of silver marked a river's course, appeared impossibly deep. Cárdenas and the others were the first Europeans to see the Grand Canyon of the Colorado River.

Instead of marveling at the ingenuity of the Native American pueblo dwellings as later explorers were to do, the Spanish raged because the cities were not made of gold. This photo was taken by an American government expedition in 1879.

CHAPTER 8

Ptolemy's Great Southern Continent

Of all the myths, legends, and mistakes that have led geographers and explorers down false trails, none lasted longer than the notion of a giant landmass in the Southern Hemisphere. That idea was born in the second century—perhaps earlier—and it persisted for sixteen hundred years, until the eighteenth century.

The ancients believed that a vast continent must exist somewhere south of the equator to "balance" the combined weight of Europe, Asia, and Africa in the Northern Hemisphere. Without such a continental counterweight, they thought, the earth would tip over and go rolling wildly away among the stars.

In the second century A.D., a Greek geographer named Claudius Ptolemaeus created a map of the world based on this idea. Ptolemy, as he was called, lived in Egypt and was familiar with the geography of North Africa. He also had a rough notion of the general shape of Asia as far as the Malay Peninsula, but he knew nothing of southern Africa or eastern Asia. On his map he joined the bottoms of both of these continents to a huge landmass that covered the whole southern rim of the world, making the Indian Ocean into a large, landlocked inland sea. Ptolemy called the southern landmass the Terra Australis Incognita, or "Unknown Southern Land." In the centuries that

James Cook, a skilled navigator and mapmaker, was the man chosen by the British government to solve the riddle of the Great Southern Continent.

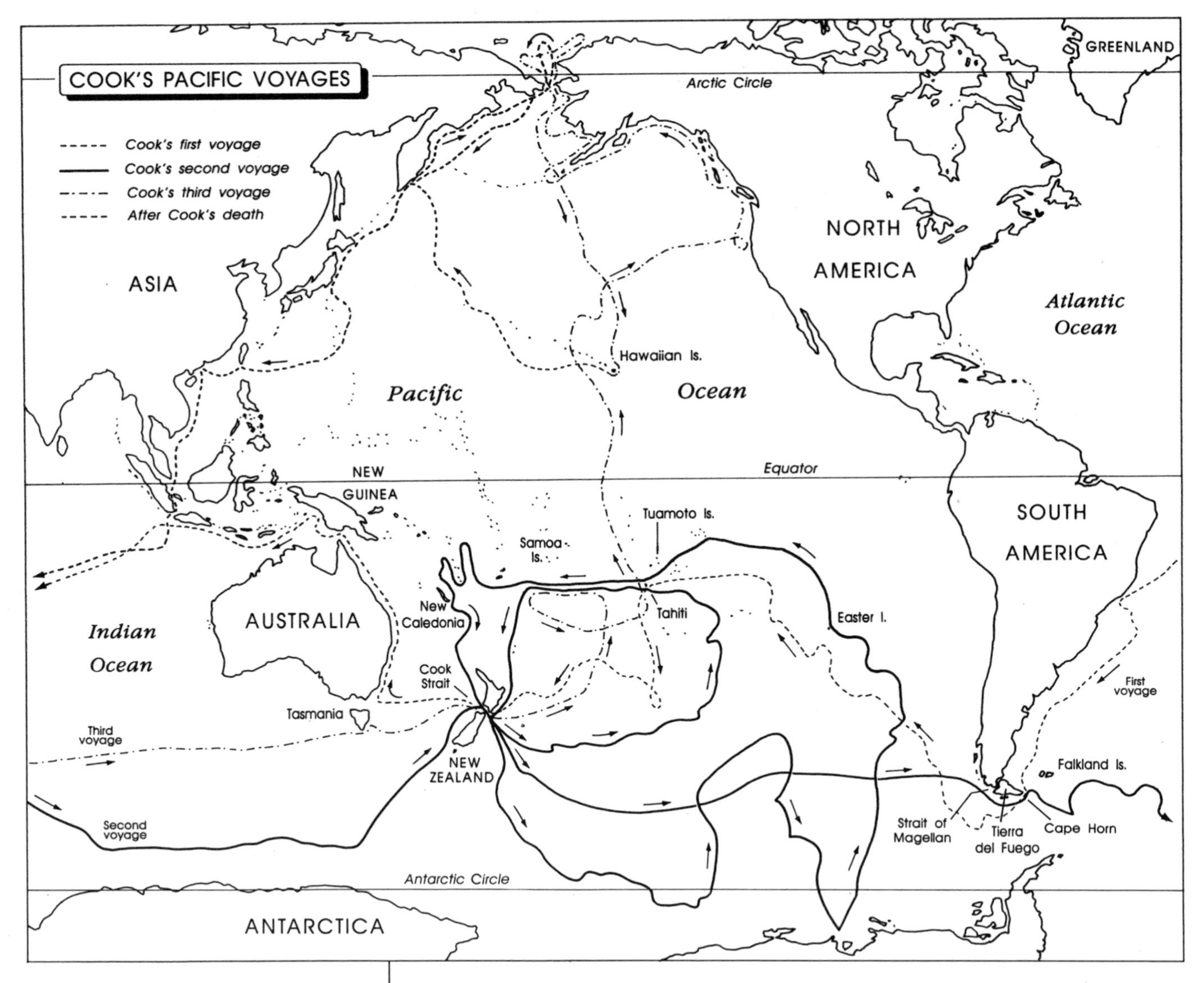

followed, the idea of an unknown southland became part of general geographical lore, even though it was based purely on speculation and not on observation or experience.

Ptolemy's vision of the world gained new popularity in Europe after 1410, when his eight-volume *Guide to Geography* was published in Latin. Little more than a century later, however, European explorers had sailed around Africa and also into the Indian Ocean from the Pacific, proving that the Indian Ocean is not a landlocked sea. Yet no one questioned Ptolemy's main idea: that a big continent lurked somewhere in the south, waiting to be discovered.

It was quickly determined that the tapering southern ends of Africa and South America together did not equal a sufficiently large landmass below the equator to make up the Terra Australis Incognita. Therefore, the Terra Australis must lie farther south, somewhere below Africa, South America, and Asia. Every time an explorer sighted land down there, he assumed it must be part of Ptolemy's great southern continent.

Ferdinand Magellan, who commanded the Spanish fleet that made the first trip around the world in 1519-22, was the first navigator to pass through the channel between the southern tip of South America and Tierra del Fuego, the island just south of that continent. The passage is called the Strait of Magellan in his honor. He did not realize that Tierra del Fuego is really a fairly small island; he thought it was the northern edge of the Terra Australis, and so did everyone else. A map made by Rumold Mercator in 1587 shows the Terra Australis as an enormous blank continent covering most of the earth south of the Tropic of Capricorn, barely a stone's throw from South America, Africa, and the Indies.

The seventeenth century brought scores of European vessels into the waters around Asia and the South Pacific Ocean. In 1616 the Dutch merchant Dirck Hartog made landfall on the coast of an unknown land south of the Indies. At last, the Europeans believed, the Terra Australis had been found. In reality, Hartog had found the island continent known today as Australia—which, while large and abundantly interesting, is nowhere near the size of the southland that had been conjectured for so many years.

In the following decades, bits of Australia's shores were mapped by the Dutch, who established a trading colony in Indonesia, a large island group north of Australia. In 1642 the Dutch sent Abel Tasman to make a more thorough exploration. Tasman sailed around Australia but did not see the entire coastline. He also made the first European landfalls at Tasmania, off the south coast of Australia, and New Zealand, far to the east. These places were naturally assumed to be part of the elusive southern continent, although they were later found to be islands. Tasman sailed again in 1644 and this time explored Australia's northern coast.

The Dutch governor of Indonesia, who had sponsored Tasman's voyages, was furious with Tasman for not bringing back a more complete and encouraging report on the Great Southland. He declared that he would have the whole region explored again "by more vigilant and courageous persons," but he died before he could do anything about it. As for Tasman, he settled down quietly in Indonesia and grew rich in the spice trade. It was left for the British to settle the mystery of the Terra Australis.

In 1767 a British geographer named Alexander Dalrymple published a book called *Account of the Discoveries made in the South Pacifick Ocean,* which contained a long and detailed argument for the existence of the Great Southern Continent. His argument was

Australia, home of the kangaroo, was long thought to be part of a large, unknown southern landmass.

much the same as Ptolemy's had been sixteen hundred years earlier. He said that a landmass "was wanting on the South of the Equator to counterpoise the land to the North, and to maintain the equilibrium necessary for the Earth's motion." He also claimed that the still-undiscovered continent must be bigger than "the whole civilized part of Asia"—big enough, he optimistically added, to hold fifty million British colonists.

Dalrymple, an armchair traveler, offered to lead an expedition to this appealing southland, but to his disgust an experienced naval officer was chosen instead.

That officer was Lieutenant James Cook, who in 1768 was assigned to take a group of scientists to Tahiti, an island in the South Pacific that had first been visited by Europeans the previous year. The scientists' purpose was to make astronomical observations—but Cook was handed a sealed packet of orders and told to open them when the scientists had finished. His secret mission was to scour the southern ocean in search of a landmass. If Dalrymple's big continent existed, the British were determined to claim it before their French rivals could do so. And if, as some suspected, it did *not* exist, they wanted the question answered once and for all.

Cook sailed west around the tip of South America and reached Tahiti in 1769. The scientists observed a rare astronomical event, the planet Venus crossing in front of the sun. They had hoped that measuring this transit of Venus would help them determine the distance from the earth to the sun, but the results were disappointingly vague. Cook was not distressed, however, because the real reason for his voyage lay ahead. He left Tahiti—to the accompaniment of tears from some of his sailors who had grown attached to the beautiful island and its amiable women—and headed south into the unknown.

He went to latitude 40° S without sighting a trace of a continent. Then he turned west and came upon New Zealand. Always a careful maker of maps and sea charts, Cook spent six months mapping the entire New Zealand coastline, proving that New Zealand was an island group, not a continent. He then continued west and came upon the east coast of Australia; he was the first European to arrive there. He also discovered eastern Australia's 1,250-mile-long (2,000-kilometer-long) Great Barrier Reef and nearly perished when his ship ran aground on its sharp coral. He and his crew patched up the ship and returned home by way of South Africa, reaching England in 1771.

The scientific and navigational knowledge that Cook brought back was exceedingly valuable, but the question of the Great Southern Continent could not yet be regarded as settled. Perhaps the landmass lay farther south than Cook had gone—much of Europe and North America, after all, was north of 40° N latitude. The British Admiralty decided to send Cook out again, this time with the sole purpose of proving or disproving the existence of the Terra Australis. To do that he would have to reach the southernmost latitude possible and then sail all the way around the world at that latitude.

At this point Cook was engaged in what historian Daniel J. Boorstin has called "negative discovery"—proving that something is not there. In *The Discoverers,* Boorstin points out that negative discovery is in some ways "far more exacting and exhausting than to succeed in finding a known objective." In other words, if you are looking for something that really exists, all you have to do is keep looking until you find it. But if you are looking for something that does not exist, you have to look everywhere that it might possibly be until you can *prove* that it does not exist—and even then someone may come along to tell you that you did not look hard enough.

Cook, now a captain, sailed in 1772 with two ships. He sailed past Africa and kept going, to a point farther south than anyone had gone before. In December, the height of the southern summer, he encountered what he described in his ship's log as "an immense field of ice, to which we could see no end." Working his way cautiously along the ice field while his sailors piled on their extra clothes, Cook crossed the Antarctic Circle at 66°32' S. Soon "the ice was so thick that we could see no further." Cook retreated to warmer latitudes and sailed among the Pacific Islands for six months.

Cook had explored the southern ocean between Africa and Australia and found it empty of all but ice. During his second southern summer, from November 1773 to March 1774, he explored the region between Australia and South America, with the same result. In January 1774 his ship halted at the southernmost point that had been reached by any explorer: 71°10' S latitude, only 1,250 miles (2,000 kilometers) from the South Pole.

Cook himself did not reach the absolute farthest south. That honor went to a sixteen-year-old sailor named George Vancouver, who clambered over the bows and clung to the ship's bowsprit in order to go farther south than anyone. (In later years Vancouver was

An account of Cook's first voyage was published in 1774. In an age of bustling expansion, the reading public was eager for books about new geographical discoveries.

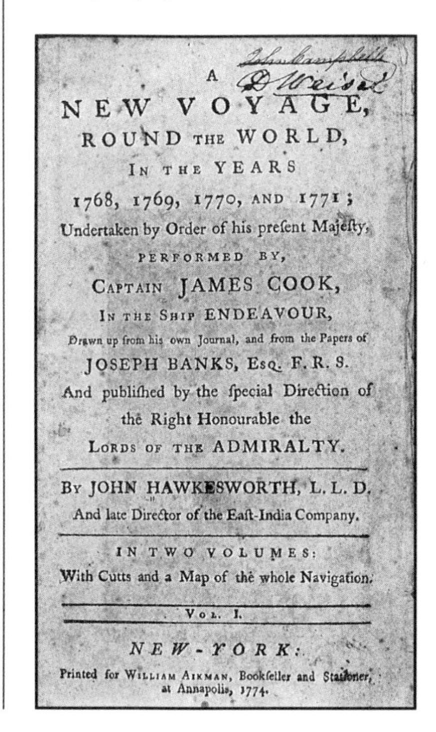
A NEW VOYAGE, ROUND THE WORLD, IN THE YEARS 1768, 1769, 1770, AND 1771; Undertaken by Order of his present Majesty, PERFORMED BY, CAPTAIN JAMES COOK, IN THE SHIP ENDEAVOUR, Drawn up from his own Journal, and from the Papers of JOSEPH BANKS, ESQ. F. R. S. And published by the special Direction of the Right Honourable the LORDS OF THE ADMIRALTY.

BY JOHN HAWKESWORTH, L. L. D. And late Director of the East-India Company.

IN TWO VOLUMES: With Cutts and a Map of the whole Navigation.

VOL. I.

NEW-YORK: Printed for WILLIAM AIKMAN, Bookseller and Stationer, at Annapolis, 1774.

renowned as an explorer of the American northwest coast; the Canadian city of Vancouver is named for him.) Cook then counted ninety-seven icebergs ahead and was forced to turn back.

Once again Cook spent the cold months of the year in the Pacific Islands. He visited Easter Island and marveled at its strange, brooding monoliths—tall stone statues of solemn faces that were erected by the island's early inhabitants. When warm weather came he turned south for a final sweep of the high latitudes between South America and Africa. Once again he found no trace of Ptolemy's Great Southern Continent.

Upon returning to England in 1775 Cook claimed, "I have now made the circuit of the Southern Ocean in a high Latitude and traversed it in such a manner as to leave not the least room for the Possibility of there being a continent, unless near the Pole and out of reach of Navigation." Cook was right about two things. First, he had finally disproved the myth of the Terra Australis. Second, there *was* a continent near the Pole, although Cook had not seen it and merely guessed at its existence. That continent is Antarctica. Cook was wrong about one thing, though. Antarctica did not long remain "out of reach of Navigation." Exploration of it began less than fifty years later.

Cook made a third voyage, this time in search of another long-lived geographical myth—the Northwest Passage. His mission was to explore the northwestern corner of North America, near the Arctic

During Cook's first visit, the people of the Sandwich Islands—now called Hawaii—treated Cook with respect and hospitality. On a later visit, however, a fight broke out between the English and the Hawaiians, and Cook was killed.

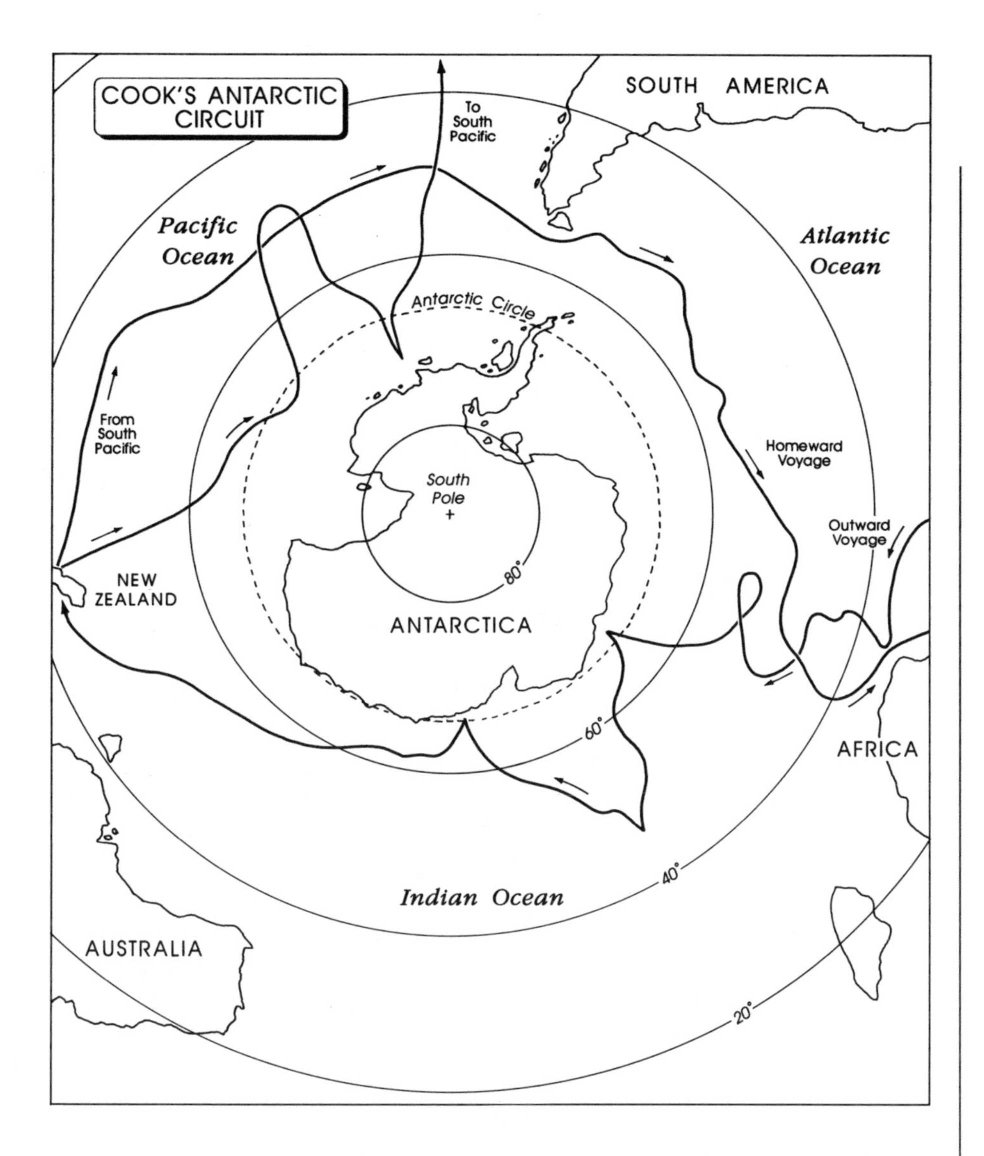

Sea, and discover whether there was a passage through or above North America to the North Atlantic Ocean.

Leaving England, he went east around Africa and across the Indian Ocean, past Australia and then north through the Pacific Ocean, toward the American northwest. On the way, in 1778, he made the first European landfall in the Hawaiian Islands. The following year, on his second visit to Hawaii, he was killed in a fight with the Hawaiian islanders.

Today Cook is highly regarded for his excellent seamanship, his meticulous and informative journals and maps, and his thoughtful and enlightened care of the men in his command. Historians of discovery call him one of the most admirable explorers of all time—although he never did find the Terra Australis.

CHAPTER 9

Northwest Passage

The Northwest Passage is not a geographical phantom like Prester John's kingdom, or El Dorado, or the Terra Australis. It does exist—after a fashion. But the real Northwest Passage is nothing like the vision that drew many explorers to their deaths in the cold, dark Arctic. In *The Oxford History of the American People,* the historian Samuel Eliot Morison wrote:

> America was discovered accidentally by a great seaman who was looking for something else; when discovered it was not wanted; and most of the exploration for the next fifty years was done in the hope of getting through or around it.

Indeed, no sooner had the European explorers stumbled on the Americas than they began looking for a way to get past them, still driven by the urge to reach the Indies. The Spanish tried to find a passage through or around Central or South America; the British and French concentrated on North America.

Just five years after Columbus's first voyage to the New World, an Italian navigator named John Cabot, sailing from England under a British flag, undertook a voyage westward in a northerly latitude. He reached Newfoundland, the island that had been briefly settled by the Vikings five hundred years before, and mistakenly believed

The English mariner Henry Hudson (opposite) was one of many who sought the elusive Northwest Passage. In 1611 his crew mutinied and set him adrift in a small boat, along with his young son and a few loyal crewmen, in the Canadian bay that now bears his name.

he had found the northeastern coast of China. He set out from England in 1498 on a second voyage and was lost at sea.

But his voyages had planted the notion of seeking a water route to Asia across the North Atlantic Ocean. The notion was reinforced by the French explorer Jacques Cartier's voyages to Newfoundland and the Gulf of St. Lawrence in the years from 1535 to 1542; for a time it was thought that the St. Lawrence River might be a sea passage to China.

Finding a northern seaway to Asia became a passion for Martin Frobisher, a sixteenth-century Englishman who spent fifteen years trying to get support for an expedition to discover what he called the Northwest Passage. Frobisher's friend George Best later wrote that the discovery of the Northwest Passage was, in Frobisher's view, "the only thing of the world that was yet left undone whereby a notable mind might be made famous and fortunate."

Finally, in 1576, Frobisher was able to seek his dream of fame and fortune. He sailed west past Greenland and came upon a wide stretch of water between "two mainlands or continents." He named this waterway Frobisher's Strait, believing that, like the Strait of Magellan at the other end of the Americas, it would prove to be a passage into the Pacific Ocean. It was not a strait, however; it was simply a long, wide bay, although that was not known for many years. It is on the southeastern side of Baffin Island, in the Canadian Arctic, and today it is called Frobisher Bay.

Frobisher was distracted from his search for the Northwest Passage by fights with the inhabitants of Baffin Island, the Native American Inuit people, who were called Eskimos by some later European explorers. An even more urgent distraction was the discovery of a lump of black stone that seemed to contain metal ore. When he turned back to England, Frobisher decided to take the stone with him. It was a fateful decision.

Martin Frobisher thought he had discovered both the Northwest Passage and an inexhaustible source of gold.

Upon reaching England, Frobisher announced proudly that he had found the passage to the Pacific. But the excitement caused by this announcement was quickly eclipsed by the news that a chemist had found a speck of gold in the mysterious black stone. The merest hint of gold was enough to drive all thoughts of exploration out of Frobisher's head. He hastily assembled funds for a second expedition, including a substantial sum from Queen Elizabeth I. Rather grandly calling himself "high admiral of all seas and waters, countries, lands, and isles, as well of Cathay as of all other countries and places of new discovery," Frobisher spent the summer of 1577 back on Baffin Island, loading his three ships with two hundred tons of rocks.

He carried the rocks back to England, where a team of chemists declared that they contained profitable amounts of gold and silver ore. At once the queen and many others bought shares in Frobisher's next expedition, and with fifteen ships he left England for Frobisher's Strait in May 1578.

After a wrong turn, in which he led his fleet into the strait leading to as-yet-unexplored Hudson Bay, Frobisher arrived at the site of his earlier excavations. He set up an efficient mining operation on that barren shore and left for home a few months later with 1,350 tons of rock. Alas for poor Frobisher—smelters spent five years trying to get gold or silver from his rocks, but failed. The tons of rock that Frobisher had lugged home from his second and third voyages were iron pyrite, aptly known as "fool's gold." It glitters, but it is not gold.

To this day no one knows whether the chemists who assayed Frobisher's first two samples were dishonest or simply incompetent. The disappointing rocks were eventually used for paving stones or

Sir John Franklin hoped to find the Northwest Passage. He failed, but his death led to the thorough exploration of the Canadian Arctic.

to shore up harbor walls, and the navigator himself became generally unpopular, especially with those who had invested in his scheme. He returned to the sea as a privateer, raiding the Spanish West Indies along with Sir Francis Drake. He later commanded a navy ship in battle against the Spanish, recovered his former glory, and was rewarded with a knighthood. He died in 1594 of a wound inflicted by a Spanish bullet.

Despite the failure of the mining venture, the Northwest Passage still acted as a magnet, drawing explorers into a bleak and lonely land. A decade after Frobisher, John Davis made three voyages into the waterway that is now called the Davis Strait, between Greenland and Baffin Island, and believed he had found the passage. He then went adventuring in the Pacific—though not by way of the Northwest Passage—and was killed by Japanese pirates. Henry Hudson also thought he had found the passage, but his discovery was really a dead end: the huge Canadian bay that now bears his name. He was abandoned there by his mutinous crew in 1611.

In the years that followed, explorers gradually pieced together a map of the maze of islands and ice-choked channels above Canada's vast and empty northern shore. The British government became so eager to find the Northwest Passage that King George III offered a large cash reward for its discovery.

A Northwest Passage would offer several advantages. It would give Britain a swift and direct trade route to Japan and China, saving ships the long journey around Cape Horn in South America or the Cape of Good Hope in Africa. It would also enable the British to keep an eye on the Russian traders and fur trappers who were moving into Alaska in increasing numbers. Eventually the British Admiralty began offering cash prizes for each new measure of westward exploration. A number of explorers were able to penetrate partway into the Arctic Archipelago, as the island maze is called; but all of them either died or were forced to turn back when their food ran out or their ships became frozen fast in the ice.

Sir John Franklin's tragic expedition was the turning point in the history of the Northwest Passage. Franklin sailed from London in 1845 with 129 men and two ships, the *Erebus* and the *Terror*. When 1846 and then 1847 passed without word from him, the British Admiralty made rescue plans. A search began in 1848 and lasted for many years, urged on by the Admiralty and Franklin's frantic wife.

Franklin had taken his ships west into Lancaster Sound at the top of Baffin Island. The expedition spent the first winter there, then, when the ice broke up in the spring, went south along Peel Sound and Franklin Strait. Winter came again, and the ships were frozen in Victoria Strait, near King William Island.

The mass of ice that held the ships was so huge that it did not break up during the following summer. Franklin died there in June 1847. His men spent a third winter in the ships, which were slowly being crushed by the ice. In the spring, knowing they would never be able to sail away, they abandoned the *Erebus* and the *Terror* and fled on foot over the ice pack, hoping to reach some distant outpost of civilization on the Canadian mainland.

The search for the Franklin expedition was a massive manhunt, involving dozens of ships and hundreds of men. But no trace of the missing ships or men was found until 1854, when Dr. John Rae of the Hudson's Bay Company recovered some silverware bearing Franklin's initials from natives near the Boothia Peninsula.

Francis McClintock, a British naval officer and explorer, carried the Franklin search to that area, and in 1859 he heard rumors among the natives along Canada's north coast. They spoke of seeing ships trapped in the ice and sinking, and of white men staggering across the tundra. "They fell down and died as they walked along," said one native woman.

Shaken by this dire news, McClintock pressed on and discovered a grim relic: a sledge and two skeletons. On the sledge was a lifeboat, loaded with all sorts of gear from Franklin's ships, including monogrammed silver dinner plates, popular novels, sheets of lead that had been used to reinforce the ships' hulls, and towels. The skeletons were those of Franklin's crewmen, who for some reason were apparently unwilling to discard all these useless items salvaged from the ships. Weakened by disease and starvation, the men were unable to pull the fifteen-hundred-pound load and died where they collapsed. McClintock also found written records describing Franklin's death. Later more bodies were found in the wilderness. Some had been buried by their shipmates; others had been gnawed by animals. No one from the Franklin expedition survived.

Franklin did not solve the mystery of the Northwest Passage, but his death led to the solution. In the course of the wide-ranging rescue effort, search parties explored all parts of the archipelago and the

The Norwegian explorer Roald Amundsen (far left) and his crew were the first to sail from one end of the Northwest Passage to the other. The trip took three years. Amundsen's next achievement was at the other end of the world, in Antarctica; he was the first person to reach the South Pole.

Canadian coast. By the time of McClintock's discovery, most of the blank places on the Arctic map had been filled in. It became clear that there *is* a sea passage from the North Atlantic to the North Pacific, twisting and turning among the islands and often blocked by solid ice.

Still, no one had yet sailed from one end of the passage to the other. Early in the twentieth century, Norwegian explorer Roald Amundsen attempted it in his little ship *Gjoa,* which was specially fortified to survive several winters in the ice pack. Amundsen sailed into Lancaster Sound in 1903 and, after three winters in the ice, sailed out into the Bering Strait on the other side of the continent in 1906, the first mariner to conquer the Northwest Passage.

Since Amundsen's time, a few navy vessels have navigated the passage. Oil companies experimented briefly with sending tankers along it, and then decided to build pipelines instead. And even one or two luxury cruise liners, icebreakers equipped with the latest safety gear, have carried wealthy thrill-seekers through the passage—something that would no doubt have astonished Amundsen. But the Northwest Passage of Martin Frobisher's dreams, that broad and navigable pathway to the Pacific, does not exist.

UNEXPECTED EXPLORERS

The word *explorer* suggests a keen-eyed, purposeful voyager or traveler, boldly venturing into the unknown to seek some geographical goal or push back the frontiers of knowledge. Yet many of the people who are honored as discoverers left home for other reasons, and found themselves exploring along the way. And sometimes specific missions of discovery, with well-known goals, fell by chance or accident to the lot of individuals who never dreamed of becoming explorers.

CHAPTER 10

Early Chinese Travelers

In 138 B.C., when the Roman Empire consisted of half a dozen Mediterranean provinces and the Romans were just beginning to extend their conquests into the Middle East, a government official named Chang Ch'ien led a party of one hundred travelers westward out of China. He carried the shaggy black tail of a yak, a large buffalolike animal native to the Himalaya Mountains; this ceremonial emblem identified him as an envoy of Wu Ti, emperor of China. Chang Ch'ien was about to travel across a region larger than the entire Roman realm.

At that time China was under periodic attack from a nomadic people called the Hsiung-nu, who came from the steppes of Mongolia. Chang Ch'ien had been sent on a diplomatic mission. The emperor ordered him to cross central Asia—thousands of miles of desert, mountains, and bandit-infested trails, almost entirely unknown to the people of the Middle Kingdom, as the Chinese called their empire. Chang Ch'ien's goal was to visit a people called the Yüeh-chih, of whom the Chinese had heard rumors from traders. The Yüeh-chih lived southeast of the Aral Sea, in a land called Bactria; their capital was Samarkand. Emperor Wu Ti wanted Chang Ch'ien to make an alliance with the Yüeh-chih against the Hsiung-nu.

Chang Ch'ien's mission could not have gotten off to a worse start. As soon as he had crossed the fortified border of China, he was captured by the Hsiung-nu. They held him prisoner for ten years, although he managed to hang onto his yak-tail emblem for the duration. He also married a Hsiung-nu woman and fathered a child.

China's borders are guarded by bleak deserts in the northwest and by towering cliffs (opposite) in the southwest. Some early Chinese rulers were as determined to keep their subjects from traveling abroad as they were to keep foreigners from entering the Middle Kingdom, as China was called. Nonetheless, a few intrepid Chinese travelers managed to roam the world.

Because the Hsiung-nu were wanderers, Chang Ch'ien saw a great deal of Mongolia, the Gobi Desert, and the steppes of Siberia during this time—probably much more than he wanted to see.

After a decade of captivity, he and some of his men escaped. Instead of trying to go home, they dutifully continued their interrupted mission. Making their way from oasis to oasis, they went west across the Takla Makan, a huge desert north of Tibet (today it is part of China). After many adventures, Chang Ch'ien arrived in the land of the Yüeh-chih, only to find that they were not at all interested in an alliance with the Middle Kingdom. He spent a year there, trying to get someone to take him seriously, and then he started for home.

More bad luck: he was captured again by the Hsiung-nu. This time, however, he escaped more quickly. He reached China after a thirteen-year absence. With him were his wife, a servant, and two members of his original escort.

Chang Ch'ien failed to make a treaty with the Yüeh-chih, but his trip brought China other valuable rewards. He was probably the first Chinese to travel so far west, and he returned with a wealth of new and useful information about the Hsiung-nu, the Gobi Desert of Mongolia, the Takla Makan, the Yüeh-chih, and the kingdoms on the far side of Asia. He was able to give reliable accounts of India and other lands he had not visited but had discussed with fellow travelers he met in Samarkand. He also brought rumors of countries still more remote—lands such as Syria, on the fringe of the Western world, around the Mediterranean Sea. He had even heard something about a distant kingdom in the West; this was probably the first rumor of the Roman Empire to reach China.

A Chinese scroll tells the story of a princess who was kidnapped by the Hsiung-nu, the nomads on China's northern border.

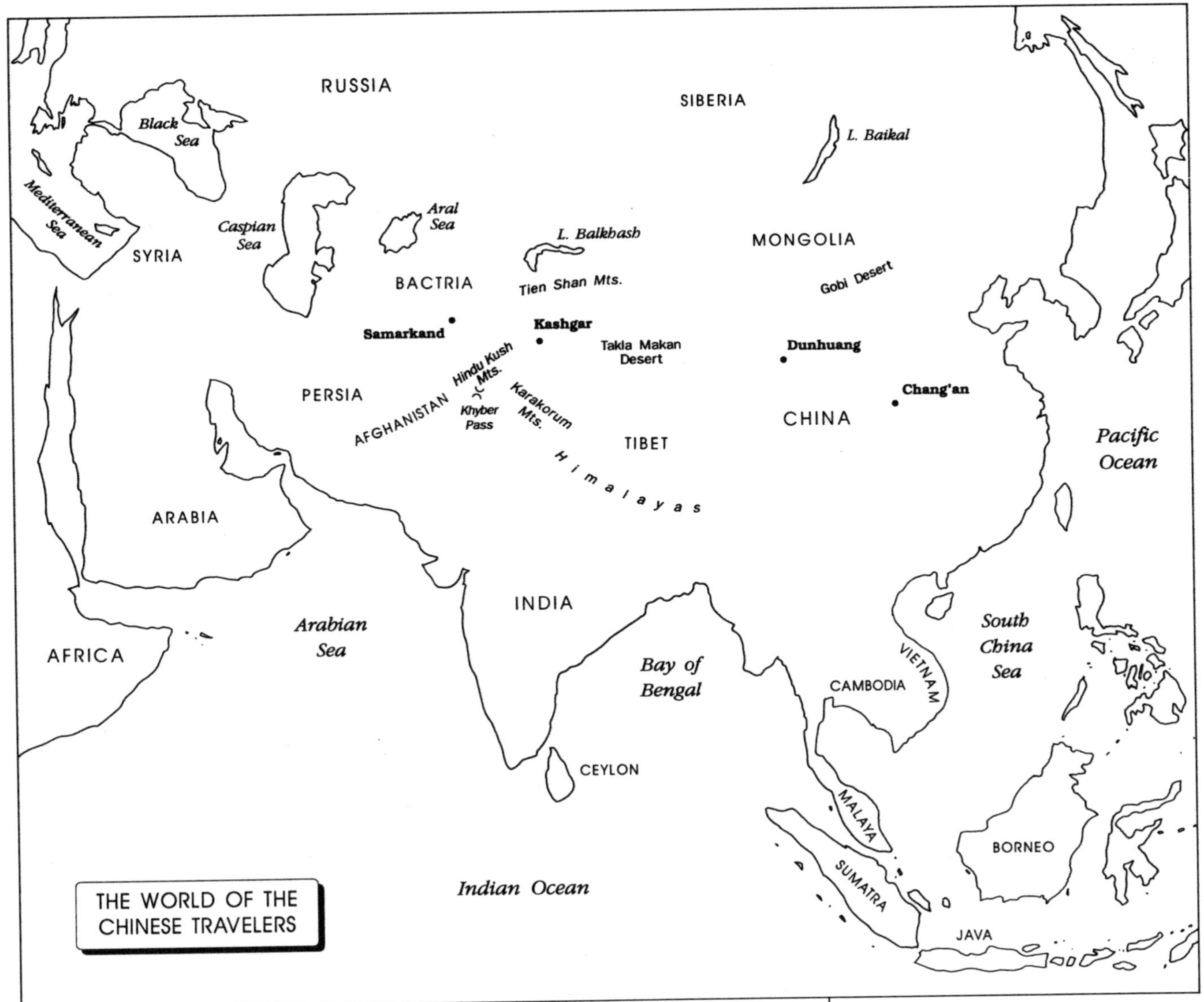

Chang Ch'ien's travels greatly enlarged China's knowledge of the world. As a result of his diplomatic mission and another, later trip he made to central Asia, China and Persia exchanged ambassadors in 110 B.C. This was the first contact between the two powers that ruled the opposite ends of Asia, and it led to increased traffic between them along the caravan route that came to be called the Silk Road. Chang Ch'ien helped to start trade between China and central Asia—the Chinese were particularly eager to trade for the fast, beautiful horses that were bred by the steppe-dwellers around Bactria. He also introduced new agricultural products, including grapes and alfalfa, that he had encountered in the oases of the Takla Makan and the plains of Bactria, and these soon became widely distributed within the Middle Kingdom. Chang Ch'ien therefore changed not only China's picture of the world but also the face of China itself.

The motives of Fahsien, a Chinese traveler of the early fifth century A.D., were neither political nor mercantile. He was a Buddhist whose fifteen-year journey was a religious pilgrimage.

Fahsien wanted to learn more about the origins of Buddhism and the extent of that faith in Asia. By his time China had entered an era of internal strife that historians call the Six Dynasties period. Six royal families ruled different parts of China and fought with one another for supreme power. The only thing that the Chinese emperors of this period agreed on was their desire to keep China cut off from the rest of the world—and therefore free of outside influences that might weaken their hold on the people. For this reason, travelers from other countries were not permitted to enter China, and the Chinese people were discouraged from leaving their country. The only Chinese who were allowed to travel abroad were religious pilgrims, and their journeys were solitary, difficult, and slow.

Fahsien set out from China in 399 or 400 and spent several years wandering through central Asia, recording the location of Buddhist monasteries. He found that there were many pilgrims on the road; he wrote that in Kashgar, an oasis city in the western part of the Takla Makan, pilgrims "collect together like clouds."

In 402 Fahsien crossed the steep and inhospitable Hindu Kush Mountains and went south into India, where Buddhism had arisen in the sixth century B.C. In Fahsien's time, India was the center of Buddhist learning and activity and the stronghold of the Buddhist faith. It was also regarded by the Chinese as a land of mystery, because there had been no trade or travel between the two nations for several centuries. Fahsien had set out on his pilgrimage in order to study and to deepen his religious faith, not to see marvels or visit faraway places, but his wanderings from one monastery to another carried him into many regions that were previously unknown to his fellow Chinese.

Fahsien went as far south and west as Afghanistan, crossing the high Khyber Pass from Pakistan into country that no other Chinese had visited. Turning back to the east, he made his way back into India, where he spent many years copying sacred texts and walking roads that the Buddha himself was said to have walked. Eventually he moved on to Ceylon, and then to Java, an island in what is now Indonesia. In Java he found himself once again on the edge of territory that was familiar to Chinese sea captains, and he made his way by sea—probably carried by a trading ship—back to his homeland. He reached China in 414 and spent the rest of his life translating into Chinese the sacred texts he had collected.

Fahsien's account of his pilgrimage is a unique record of Buddhism during its most flourishing era, before Islam arose in Arabia to challenge Buddhism's strength in central Asia. But Fahsien's story is also a vivid and colorful travel narrative that gave learned Chinese Buddhists a glimpse of the busy, variegated world outside the Middle Kingdom that most of them would never see.

Two centuries later, in 629, another Buddhist pilgrim set out from China. He was Hsüan-tsang, a scholar and minor aristocrat who wanted to "travel in the countries of the west in order to question the wise men on points that were troubling his mind," as a contemporary biographer explained.

A few years earlier, China had come under the control of the Tang Dynasty, which was to rule the country until 907. In later years the Tang rulers expanded the Chinese Empire to its greatest extent and took a deep interest in the rest of the world. But in the first years of the Tang Dynasty, China was even more isolated from the world than it had been during Fahsien's lifetime. All Chinese people were now completely forbidden to travel, so Hsüan-tsang had to sneak past the watchtowers on the western roads. He was caught once, but the guards were Buddhists and they sympathetically let him go.

Hsüan-tsang's route was much like Fahsien's, but his luck was more like Chang Ch'ien's. Soon after entering the desert he dropped his water bag, and he would have died of thirst if his horse had not scented the way to the nearest oasis. He worked his way around the northern edge of the Takla Makan through the snowy Tien Shan Mountains, which he described as "sheets of hard, gleaming white, losing themselves in the clouds."

Hsüan-tsang crossed Asia as a Buddhist pilgrim seeking enlightenment and wisdom. Along the way he acquired a wealth of geographical knowledge. In this Japanese painting from the early ninth century he is dressed in Japanese style, with an elaborate backpack.

Having obtained safe conduct from the bandit warlord who ruled the region, Hsüan-tsang then visited Samarkand, the westernmost point that he was to reach on his travels. Like Fahsien before him, he turned south to cross the Hindu Kush on one of the ancient caravan paths that thread the towering, sprawling mountain ranges of central Asia. He came to a narrow pass called the Iron Gate; bandits had blocked the pass with a huge iron door, hung with "a multitude of little bells of iron" that jingled to alert the bandits to a traveler's passing.

Hsüan-tsang survived the perils of the mountains and lived for a time in India, walking the holy trails and studying with Buddhist teachers. In 645, after sixteen years of wandering, he returned to China, expecting to be arrested for having broken the ban on foreign travel. He learned that at first Emperor Li Shihmin had been furious with him, a former member of the imperial court who had dared to break the official ban on travel.

Fortunately, by the time of Hsüan-tsang's return, the emperor was no longer angry, merely relieved that the wanderer was still alive. China's official policy on travel was beginning to change; in addition, the emperor was planning wars of conquest in central Asia, so he was keenly interested to hear whatever Hsüan-tsang could tell him about the peoples, armies, roads, and other features of the region. Hsüan-tsang, who had feared punishment, found himself greeted as a returning hero.

Word of Hsüan-tsang's remarkable journey had spread, and the streets of Chang'an, then China's capital city, were filled with crowds who wanted to see him. The emperor invited Hsüan-tsang to the palace and listened to his story, which he ordered the royal scribes to record. The emperor also offered him a job, but Hsüan-tsang turned it down. Again like Fahsien, he spent the rest of his life translating Buddhist writings. His scholarly labors—and the public enthusiasm aroused by his long pilgrimage—contributed to a revival of Chinese Buddhism in the years following his death in 664. But like Fahsien before him, he was admired by his countrymen as much for his daring and his travels as for his faith and his scholarship.

Archaeologist Aurel Stein (center, with dog) spent nearly six decades exploring the trade routes and oases of central Asia.

Centuries later, it was to seem as though the hand of Hsüan-tsang had stretched across the years to help bring about an amazing—and accidental—discovery. Aurel Stein was a Hungarian-born archaeologist who in 1886 went to India to work for the British administration there. He loved traveling in Asia, and he devoted the rest of his life to crisscrossing the continent, exploring the old Silk Road route for clues about the trade, agriculture, art, and life that once flourished in its remote oases and forgotten, half-buried towns. Interrupted only by periods of war during which he was unable to travel, he continued his work until his death in 1943.

Stein's most important discovery was made in 1907, at a place called Dunhuang in western China. In bygone eras Dunhuang had been a prosperous and important city at the eastern, or Chinese, end of the Silk Road, where the caravan route entered China. For centuries it was also the home of many Buddhist monks, who over the years created the phenomenon called the Caves of the Thousand Buddhas. This was a valley whose steep walls contained many caves of all sizes. The Buddhists lived in some of the caves and used others as temples. Gradually the caves became filled with images of the Buddha, both statues and wall paintings, large and small. Some caves were nothing more than tiny niches in the rock, in which a small statue of the Buddha had been roughly carved into the cave wall. Others were immense caverns occupied by huge standing or sitting statues, or coated with plaster and painted everywhere with hundreds of sacred images.

Stein did not discover the Caves of the Thousand Buddhas; they had always been known to the Buddhist population of central Asia and had been visited by several Western explorers and scholars before Stein came to Dunhuang. None of these earlier visitors, however, had discovered what lay behind a crack in the plaster wall of one cave. One day a priest named Wang Tao-shih, one of the handful of monks who remained in the region to tend the caves, was cleaning the wall when he noticed the crack. He probed it and discovered that it led into a secret chamber—a long-forgotten room crammed not with gold and jewels but with treasures perhaps even more precious, thousands and thousands of ancient documents, most of them in the form of rolled scrolls, dry and brittle with age. In alarm, Wang closed the secret chamber up again. He did not want anyone to know of his discovery because he was afraid that it would attract the attention of the Chinese authorities, who might force the monks to leave the caves.

An old Chinese proverb says that a secret is like an egg: once it is spilled, it will not go back into its shell. This was the case with Wang's secret. Somehow, rumors of his find drifted out into Dunhuang. It happened that Stein was camped in Dunhuang at the time, making archaeological excavations, and he heard the rumors. Determined to learn whether there really was a treasure—and, if so, to preserve it for scholarly study—Stein made haste to seek out Wang, but the monk angrily denied any knowledge of old papers.

Stein had an inspiration. He told Wang that he was a learned and pious scholar, traveling all the roads of Asia in search of knowledge about the Buddhist links between China and India in the distant past. His patron saint and hero on these travels, he said to Wang, was Hsüan-tsang, the seventh-century Buddhist pilgrim, who also had made a long pilgrimage to find and preserve Buddhist texts. These words had more effect on Wang than weeks of pleading and arguing would have had. Like all central Asian Buddhists, Wang knew the story of Hsüan-tsang and regarded the pilgrim as a saint. Touched by Stein's reference to the holy one, Wang agreed to reopen the secret room. But the monk was still afraid. The most he could do, he said, was to open the room alone that night, remove one scroll, and bring it to Stein's camp.

Serendipity was at work in Dunhuang that night. Wang did as he had promised. He reached into the chamber, plucked a scroll at random, and took it to Stein. When they examined it, they found to their amazement that it was a collection of *sutras,* or sayings of the Buddha, that had been collected in India, translated and written down in Chinese, and brought to Dunhuang by Hsüan-tsang himself, nearly thirteen hundred years before. By an astounding coincidence, Wang had unknowingly put his hand on a relic of the very man Stein had claimed as his patron saint.

This fortunate accident made a deep impression on Wang, who agreed to let Stein and his assistant visit the chamber, survey its contents, and purchase some of the scrolls and wall hangings. Stein had access to the cave for only five days, so he was unable to make a complete inventory of its contents. It was clear, however, that he had stumbled upon a find of immense archaeological importance. The cave scrolls of Dunhuang were as significant to the study of central Asian history as the Dead Sea Scrolls, discovered in a cave in Israel many years later, are to the study of biblical history.

Among other things, Stein found materials that enabled him to translate two previously unknown Asian languages; he also uncovered a scroll called the Diamond Sutra, which is thought to be the world's oldest known printed book. Printed from carved wooden blocks, it contains writing and illustrations and was made in China in 868, centuries before the printing press had been invented in Europe. Stein estimated that documents had been stored in the chamber as early as the ninth century, and that it had been sealed

The Diamond Sutra is thought to be the oldest printed book in the world. It was printed from carved blocks of wood in China in 868 and contains texts and pictures related to early Buddhism.

sometime in the eleventh century, perhaps to protect its contents from bandits. Many of the documents the chamber contained, however, were much older than the ninth century and were probably brought to Dunhuang from other sacred sites; some of them are from as far back as the third century.

In the end, Wang allowed Stein to carry away twenty-nine packing cases of documents and embroidered silk banners and hangings. But thousands upon thousands remained. A French scholar who was passing through Dunhuang the following year bought some scrolls from Wang, and Wang gave Stein six hundred or so manuscripts when Stein returned to Dunhuang some years later. This time, however, the Chinese authorities heard about the cave. The regional governor loaded the remaining manuscripts onto carts and carried them away, and they were never seen again. They are thought to have been lost, destroyed, or sold to private collectors. But thanks to Stein's lucky reliance on Hsüan-tsang to open the way, some of these ancient works—the largest cache of central Asian documents ever found—survived to be preserved and studied.

CHAPTER 11

A Mountain Man in the Mojave

One of the most important pioneers of the American West became an explorer while searching for a large, water-dwelling rodent. He was Jedediah Smith, a beaver trapper. Smith was born in New-York State in 1799 and grew up in Pennsylvania. Tradition says that a family friend gave him a book about the Lewis and Clark expedition, which made the first overland journey across the United States in 1804-06. Reading about the sweeping plains, soaring peaks, and rushing rivers of the West fired young Smith's imagination and made him yearn to see the West for himself. Fur trapping offered the way.

The furs of wild animals—and especially the sleek, waterproof pelts of beaver—were highly prized in London and around the world, and many companies were formed to export furs from the frontier regions of North America. In 1822 Smith arrived in St. Louis and joined a trapping expedition sponsored by the Rocky Mountain Fur Company. His first trip for the company took him to the upper reaches of the Missouri and Yellowstone rivers, in present-day North Dakota and Montana. Soon Smith became a leading member of the vigorous, colorful fraternity called the Mountain Men, the guides and trappers who lived and worked in the Rocky Mountains during the years when "civilization" ended at the Mississippi River.

On his first expedition into the Rocky Mountains, Jedediah Smith proved his toughness by surviving the attack of a ferocious grizzly bear.

The typical Mountain Man dressed in animal skins and was never without his rifle.

Smith's first journey into the wilderness was almost his last. In the Black Hills of South Dakota he was attacked by a grizzly bear. His companions killed the bear, but not before it had ripped Smith's scalp and one ear almost off his head. Smith remained conscious and calmly told one of the others to stitch him back together. This was done, and Smith recovered fully in ten days, although he bore a grim scar for the rest of his days.

For four years, Smith worked for the Rocky Mountain Fur Company as a field guide, leading trappers into the Rockies to collect loads of beaver skins and then back to a meeting place where company representatives would pay for the furs with clothing, ammunition, and cash. These annual gatherings came to be called Rendezvous, which is French for "meeting" or "coming together"; many of the early fur trappers were French Canadian. The Rendezvous was held each year in a different place in the western mountains and attracted hundreds of Mountain Men, traders, and Indians; it was a wild month-long party as well as an annual business event. Smith was better educated than many of the hard-living Mountain Men and stood out from them in other ways as well. He prayed regularly, he seldom drank, he sent money home to his parents, and he did not smoke or swear. But he was universally respected for his skills as a woodsman and tracker, and no one ever questioned his toughness.

In 1826 Smith and two other Mountain Men, David Jackson and William Sublette, bought the Rocky Mountain Fur Company. Smith planned an expedition southwest from the Great Salt Lake in present-day Utah. He wanted to lead a party of trappers into new territory where beaver might be numerous—but as time went on he had also found himself increasingly drawn by the love of exploration for its own sake.

In a letter he wrote later to his boyhood hero General William Clark (of the Lewis and Clark expedition), Smith recalled, "I started about the 22nd of August, 1826, from the Great Salt Lake, with a party of fifteen men, for the purpose of exploring the country S.W. which was entirely unknown to me, and of which I could collect no satisfactory information from the Indians who inhabited this country." He explains his two motives this way: "I, of course, expected to find Beaver, which with us hunters is a primary object, but I was also led on by the love of novelty common to us all which is much increased by the pursuit of its gratification." He certainly found novelty. Before it was over, Smith's trapping trip included two milestones in American exploration.

Smith bore south through western Utah, passing through what is now Zion National Park. He reached the Colorado River along the present-day border between Arizona and Nevada, then worked his way through the Black Mountains in northwestern Arizona, following the river's course.

The whole route had been over flat alkali plains or high-prairie mesa, with few trees, little water—and no beaver, as the disappointed and increasingly quarrelsome trappers pointed out. Food was nonexistent, and after a few weeks men and horses alike were thin and perpetually hungry. They were also thirsty, as the only water they could find was muddy and full of insects. But in October they came out of the mountains into the long, lush green Mojave Valley

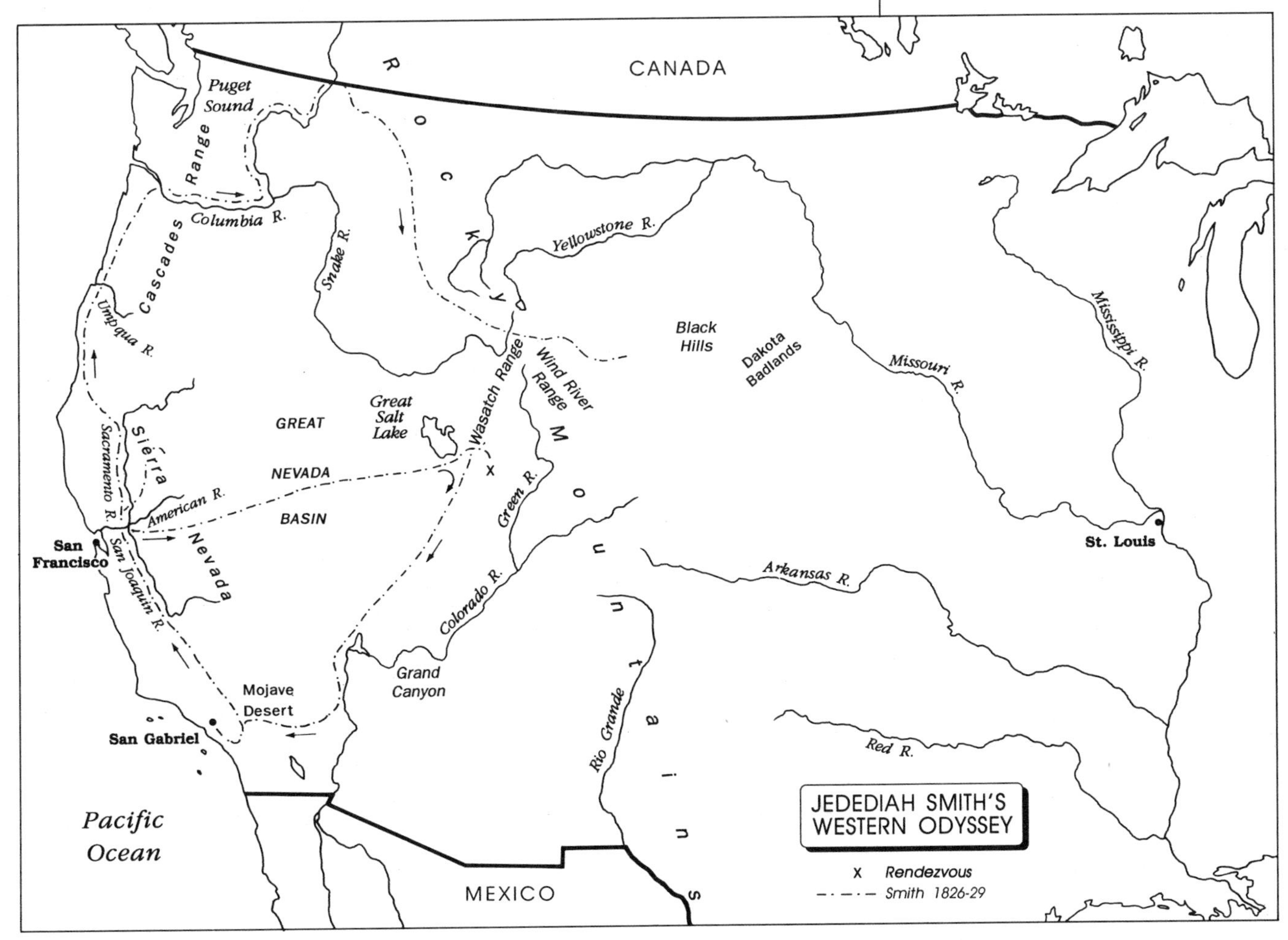

along the present-day California border. They spent two weeks there, feasting on corn and melons while they established friendly relations with the courteous Mojave Indians.

The land to the west, Smith learned from the Indians, was a desert, but the Indians knew of trails that crossed it, for they sometimes traded with other Native American peoples on the Pacific coast. They told Smith that on the other side of the desert, not far away, lay California, which was then part of the Mexican republic. There was no love lost between the people of the United States and the Mexicans, but Smith decided to try for California anyway, to complete his survey of the Southwest.

Smith's party entered the desert with two Indian guides. The shadeless expanse of sand and rock was like an oven; the men had to bury themselves neck-deep in sand to get a little relief from the burning heat. Thirst haunted them like a specter. But after fifteen days they came out on the far side of the wasteland, crossed the San Bernardino Mountains, and found themselves in a fertile valley. They had reached the Mexican settlement at San Gabriel, later called Los Angeles, and they were the first men other than Indians to cross the Mojave Desert.

Smith and his comrades found the Mojave Desert to be a desolate expanse of sun-baked sand and salt pans—places where water had dried up long ago, leaving a barren crust of mineral salts.

On the far side of the desert the trappers found the mission at San Gabriel, a lonely Spanish outpost on the site where the city of Los Angeles sprawls today.

Their troubles were not over. They were treated with great kindness by the Franciscan monks of San Gabriel, but the Mexican governor of California was not sure what to do with them. The newly created Mexican state desperately wanted to keep its Californian province free from American encroachment, but it did not want to provoke war with the United States by mistreating American citizens. The governor detained Smith for a month or so while he fretted over the dilemma, then, in January 1827, told him to go back the way he came.

Instead of returning to the Mojave Desert, however, Smith led his men north through California, past present-day Yosemite National Park. They reached the American River in May. During this journey through the foothills of the Sierra Nevadas, the trappers finally found beaver in abundance. But when they tried to go east for the summer Rendezvous in the Rockies, the snow-blocked passes and bitter cold of the high Sierras forced them to turn back. Smith set his men up in a secure camp; then, with only two companions, he tried again to cross the Sierras.

They made it through the mountains in eight grueling days, completely unaware that they were passing through a region laden with gold—the site of the California Gold Rush in 1848. Ahead of them now lay a desert even more forbidding than the Mojave: the Great Nevada Basin, 300 miles (480 kilometers) of almost waterless salt flats. With little food and water, and no friendly Indians to help them, they had only one hope: to cross the basin quickly and reach the Great Salt Lake, where the Mountain Men would be gathered.

They raced across Nevada, sometimes covering 40 miles (64 kilometers) in a single day. One of Smith's companions grew so weak that he could not go on. The other two could not carry him, so they left him stretched under a small tree and hurried on, as Smith declared, "with the hope that we might get relief and return in time to save his life." By great good fortune they found some water and were able to go back to their abandoned comrade and rescue him. They reached the end of the Great Nevada Basin in late June, and on July 3, 1827, they rode into the Mountain Men's Rendezvous. "My arrival caused considerable bustle in camp," Smith later recounted laconically, "for myself and party had been given up as lost."

Almost at once, Smith left for a second trip across the Mojave and up along the Sierras to pick up the men he had left behind in central

A beaver trapper at work. Ironically, the search for the water-dwelling beaver led Jedediah Smith across two of North America's driest deserts.

California. He reached them safely, although the Mojave Indians were no longer friendly because they had been attacked by another party of whites. Smith ran into more trouble with the Mexican authorities, but eventually he led the trappers up into Oregon. There, along the Umpqua River, they were massacred by Indians; only Smith and one other escaped. Smith himself died several years later, fighting a party of Comanche Indians on the Santa Fe Trail.

Jedediah Smith traveled in the West for less than a decade, but he saw more of it than anyone else of his time. He started as a trapper traveling to make a profit and became an explorer because he had to explore to get furs. He remained a trapper first and an explorer second throughout his short career, but he became increasingly aware of his role as a pioneer who was helping to open the way west for American settlers. He made detailed maps of his travels and kept notes in which he recorded things that might help settlers, such as information about passes, water sources, friendly and unfriendly Indians, and likely sites for farming or ranching.

Smith hoped to publish these journals and maps—perhaps the most complete storehouse of information about the West that had been compiled since the Lewis and Clark expedition—but he died before he could do so, and most of his papers were lost at his death. Nevertheless, his letters and stories of his travels circulated widely among other explorers and also among settlers, who began moving into the West just as he had hoped they would. Because of these letters, in which he shared his experiences and observations, and because he helped other Mountain Men and pioneers make their way into unexplored territory, Smith is now regarded as one of the key figures in the opening of the American West. His achievements as a trapper have been forgotten, but he is remembered as the first U.S. citizen to cross the Mojave Desert and the Great Nevada Basin.

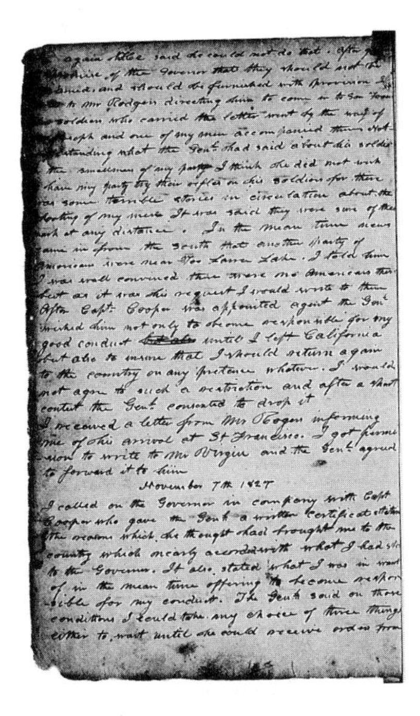
again [illegible] said he could not do that. After [illegible]
[illegible] of the Governor that they should not be
[illegible] and should be furnished with provision [illegible]
[illegible] to Mr Rodgers directing him to come in to San [illegible]
[illegible] soldier who carried the letter went by the way of
[illegible] and one of my men accompanied them Not-
withstanding what the Genl had said about his soldier
[illegible] the smallness of my party I think he did not wish
[illegible] have my party try their rifles on his soldiers for there
was some terrible stories in circulation about the
shooting of my men It was said they were sure of their
mark at any distance. In the mean time news
came in from the South that another party of
Americans were near Too Larre Lake. I told him
I was well convinced there were no Americans there
but as it was his request I would write to them
After Capt. Cooper was appointed agent the Genl
wished him not only to become responsible for my
good conduct until I left California
but also to insure that I should return again
to the country on any pretence whatever. I would
not agree to such a restriction and after a short
contest the Genl consented to drop it
I received a letter from Mr Rogers informing
me of his arrival at St Francisco. I got permission to write to Mr Virgin and the Genl agreed
to forward it to him
November 7th 1827
I called on the Governor in company with Capt
Cooper who gave the Genl a written certificate statin[g]
the reasons which he thought had brought me to the
country which nearly accorded with what I had st[ated]
to the Governor. It also stated what I was in want
of in the mean time offering to become responsible for my conduct. The Genl said on those
conditions I could take any choice of three things
either to wait until he could receive orders from

The journal Smith kept in California has been preserved, but many of his papers—filled with facts about the people and geography of the West from Canada to Mexico—were lost when he died.

CHAPTER 12

The Missionary and the Reporter

One of the most celebrated encounters in history took place on November 10, 1871, at a place called Ujiji in what is now the country of Tanzania, in East Africa. The two men who met and shook hands on the shore of Lake Tanganyika that day were among the greatest explorers of all time, although neither of them had planned it that way. One of them started out as a missionary, the other as a newspaper reporter.

The missionary was Dr. David Livingstone, the son of a poor Scottish mill worker. He seemed destined for a life of poverty and hard labor, but he determinedly completed his education. Livingstone felt the pull of scientific study and of the far horizon, and he decided to become a medical missionary in Africa. He received his doctor's degree at age twenty-seven and in 1841 was sent by the London Missionary Society to Bechuanaland, in southern Africa.

Livingstone was assigned to the mission at Kuruman—the most remote of all the British missions in that part of Africa. Yet Kuruman did not seem remote enough to Livingstone. Before long he yearned to penetrate still farther into the unknown interior of Africa that stretched away to the north of Kuruman. He felt hemmed in and confined in Kuruman, where he was supervised by an older

"Dr. Livingstone, I presume," said reporter Henry Morton Stanley in an African village in 1871 (opposite). Stanley (at left) had found the British missionary and explorer David Livingstone, whose whereabouts were a mystery to the world.

On one of Livingstone's early journeys north of Kuruman he was mauled by a lion; he never regained the full use of one arm.

missionary, and he wanted to go to a place where no European had gone before him.

Over the next decade Livingstone married the daughter of his superior at the mission, started a family, and tried several times to cross the sun-baked wastes of the Kalahari Desert, which lay north of Kuruman. He dreamed of founding a mission in the wilderness on the other side of the desert. In two attempts to cross the desert, however, he was forced to turn back; on the second attempt he was accompanied by his pregnant wife and three small children, all of whom suffered terribly from fever and thirst.

During these years Livingstone discovered that he was not very effective as a missionary. He was supposed to bring Christianity to the "unenlightened," but in all the years he spent in Africa he baptized only one convert. He became increasingly convinced that he could best serve the cause of Christ in Africa by helping to open up the yet-unexplored parts of the vast continent to British travel, trade, and settlement.

In 1851, on his third attempt, he succeeded in crossing 700 miles (1,120 kilometers) of the barren Kalahari Desert and sighted the Zambezi River flowing grandly to the east. "How glorious! How magnificent!" he cried exultantly. This was the first of Livingstone's four major explorations in Africa.

The second took place from 1853 to 1856. Once again he crossed the Kalahari to the Zambezi, and then he went westward up the river.

He wanted to explore the region between the headwaters of the Zambezi and the west coast of Africa. Though plagued by torrential rains, malaria, and demands for bribes from the native chieftains he met—he literally lost his shirt to them, but managed to keep his journals and scientific instruments—he forged ahead. Four months later, after crossing 1,500 miles (2,400 kilometers) of unmapped country, he reached the Atlantic coast at Luanda, in present-day Angola.

For many explorers, that would have been enough, but Livingstone's real desire was to be the first person to travel all the way across Africa. He had reached the Atlantic coast—but that was merely the starting point for the journey to the opposite coast. So Livingstone at once turned around and went back by a new route. It took him a year to reach the familiar banks of the Zambezi, and then in 1855 he started east along the river toward the Indian Ocean coast.

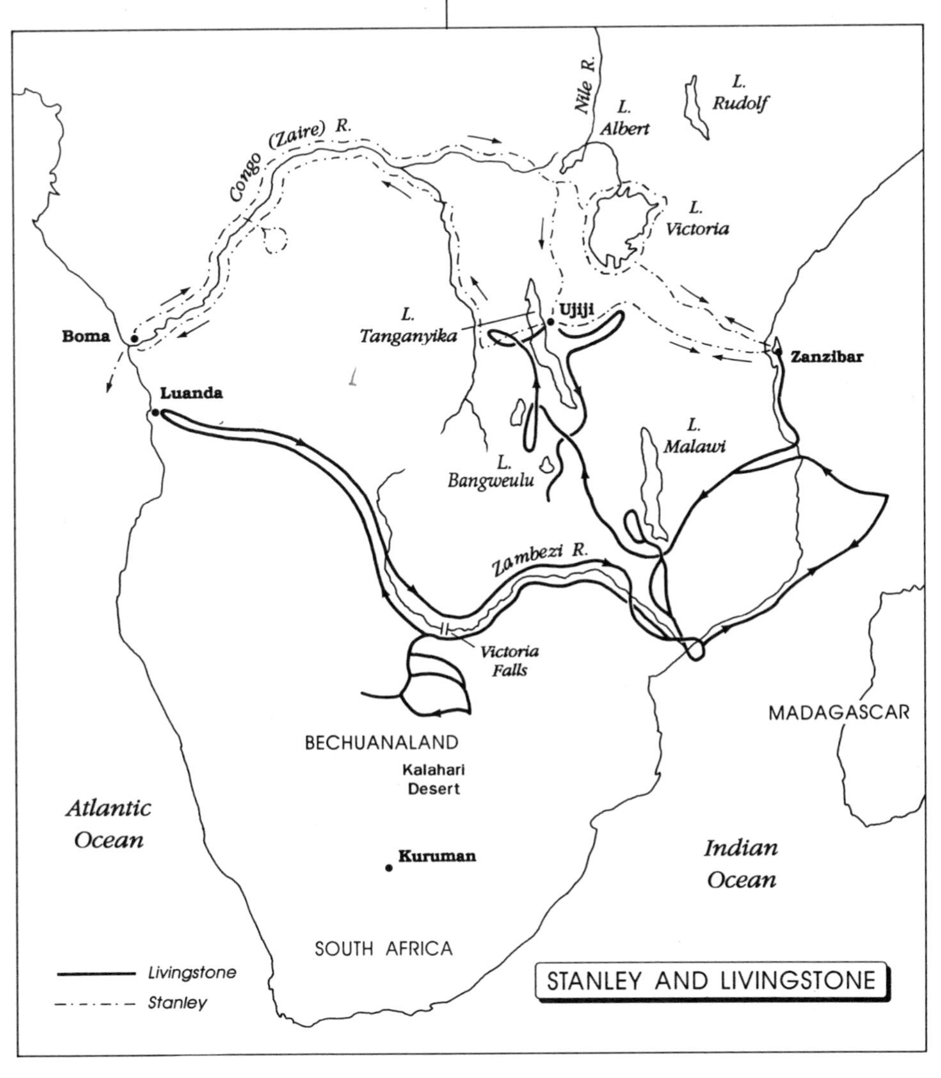

STANLEY AND LIVINGSTONE

He made the most beautiful discovery of his career along an unexplored stretch of the Zambezi. It was a deep chasm into which the river poured in a wide waterfall, called by the natives "the smoke that thunders." Livingstone named it Victoria Falls in honor of the British queen. "It had never before been seen by European eyes; but scenes so lovely must have been gazed upon by angels in their flight," he wrote.

Livingstone reached the mouth of the Zambezi, on the Indian Ocean, in 1856. He was the first European known to have traveled across Africa—a feat that becomes even more impressive when we remember that he had to go halfway across the continent in one direction before he could even begin to cross it in the other.

Upon returning to England, he was showered with awards and recognition. In nineteenth-century Europe, explorers were public heroes, and their exploits were the subject of newspaper headlines, popular songs, and endless discussion. The London Missionary Society decided that Livingstone was now an explorer, not a missionary, and it withdrew its support. However, he easily found backing in the British government for an expedition to take a steamboat up the Zambezi, which he called "God's highway into Africa." He felt that only by opening the interior of Africa to regular commerce could the slave trade be wiped out and Christianity brought in, and he wanted to show that steamboat traffic was practical.

The expedition set out in 1858 and was a dismal failure. Not far from the coast Livingstone ran into a 40-mile (64-kilometer) stretch of rapids that he had failed to see on his previous journey because his route left the river's course for a while. The rapids formed an impassable barrier, and the steamboat was halted for good. For several years Livingstone surveyed the area around Lake Malawi and gathered information on the slave trade, but the government withdrew its support in 1863 after three members of a missionary party in Livingstone's area were killed by the Africans. Livingstone's return to England in 1864 was far from triumphant. He was largely ignored by the public.

No longer a missionary, yet unable to rest in England, Livingstone returned to Africa in 1866 to search for the source of the Nile River. For years explorers had been seeking the river's source in the tangle of East African lakes and rivers. Some claimed to have found the source in Lake Victoria or Lake Tanganyika, but Livingstone believed that the true source had not yet been discovered. He wanted to be the one to locate what the ancient Greek geographer Herodotus had called "the fountains of the Nile." This was Livingstone's most ambitious expedition, and it was to be his last.

Livingstone had heard of a remote lake called Bangweulu, south of Lake Tanganyika, and he thought it might be the Nile source. He reached it in 1868 but found only a leech-infested, fever-ridden swamp. By now Livingstone was worn out and ill, and there seems little doubt that his mental state was somewhat unbalanced, at least part of the time. Susi and Chuma, two loyal African servants who had accompanied him for years, said later that he had become weak

Livingstone's African servants, Chuma (left, with a pipe) and Susi, followed the explorer in his wanderings from 1864 until his death in 1873.

and vague. Yet as he wandered between Lakes Bangweulu, Malawi, and Tanganyika over the next few years, he continued to make notes on the geography, inhabitants, and natural history of the region. He also observed the devastation wrought across the land by the cruel slave trade. "The strangest disease I have seen in this country," he wrote in his journal, "seems really to be broken-heartedness, and it attacks free men who have been captured and made slaves."

The outside world received no word of Livingstone after 1868. The fickle public forgot that his last expedition had been a disaster and remembered only that he was a hero. Now he was lost in the heart of what many called the Dark Continent. There was growing concern over his fate, and in 1870 an American newspaper editor sent reporter Henry Morton Stanley to look for him.

Stanley arrived in Zanzibar, an Indian Ocean island off the coast of East Africa, in 1871. Livingstone had been gone for five years. To find him, or carry word of his fate back to the anxious world, would be the biggest scoop the thirty-year-old newspaperman could imagine.

Stanley's background was even more humble than Livingstone's. Accounts of his origins are confusing and sometimes contradictory, but it is certain that he was born in Wales, probably illegitimately; his real name was John Rowlands. His family abandoned him in a workhouse, a sort of orphanage, and the anger he felt at this betrayal never left him. Bitter and tormented, Rowlands ran away to sea as a cabin boy when he was fifteen years old. He came to New Orleans, where he was befriended by a merchant named Henry Morton Stanley, whose name he adopted.

After the American Civil War—during which young Stanley fought on both sides—he made his way to New York and became a journalist. James Gordon Bennett, editor of the *New York Herald,* was so pleased with Stanley's work that he assigned him to search for Livingstone. Bennett's sponsorship of the search was not just charity: nothing would raise newspaper circulation more than a report that the missing explorer had been found.

Stanley left Zanzibar with ample supplies, the best equipment that the *Herald's* money could buy, and 192 African porters. He caught malaria and was entangled in battles between Arab slave dealers and Africans, but, eight months after leaving the coast, he marched into Ujiji with the American flag flying and guns firing.

A workhouse boy turned newspaperman, Stanley had a gift for vivid, lively writing. His first book, *How I Found Livingstone,* was a best-seller. Other expeditions, and other books, soon followed.

According to his later account, Stanley was astonished when he heard a voice say, “Good morning, sir!”

He turned to see a smiling African who introduced himself as Susi, Livingstone’s servant.

“Is Dr. Livingstone here!” Stanley exclaimed in surprise. Susi led the way to where the doctor waited.

Stanley described the historic meeting: “I would have run to him, only I was a coward in the presence of such a mob—would have embraced him, only, he being an Englishman, I did not know how he would receive me. So I did what cowardice and false pride suggested was the best thing—walked deliberately to him, took off my hat, and said:

“‘Dr. Livingstone, I presume.’”

Those four words became one of the world’s most-quoted sentences. The phrase quickly became famous, and later in his life Stanley often declared that he wished he had never said it; he was embarrassed because it sounded so bland and pompous.

Livingstone raised his faded cap to Stanley, and the two men shook hands. Stanley said, “I thank God, Doctor, I have been permitted to see you.”

“I feel thankful that I am here to welcome you,” Livingstone replied.

After these polite exchanges, Stanley supplied the thin, ailing Livingstone with food and medicine. “You have brought me new life!” the doctor said—but he refused to return to England or even to the coast with Stanley. He was determined to solve the riddle of the Nile. He gave Stanley letters and papers to take back; Stanley left supplies with him. Then the reporter returned to Zanzibar to write his sensational newspaper story. He later expanded it into a book, *How I Found Livingstone.*

Stanley’s career was made, and his account of Livingstone, reflecting his genuine admiration for the older man, elevated Livingstone to new heights of heroism. The image of Livingstone created by Stanley and other Victorian writers overlooked his flaws—his irritability, stubbornness, selfishness, and occasional heartlessness—in favor of the myth of the saintly missionary. Recently, however, scholars have begun to paint a picture of Livingstone’s true nature, complex and often unhappy, but more human and more fascinating than the myth.

How I Found Livingstone not only made Livingstone sound like a saint, but it made Stanley into a popular hero. Before long Stanley had taken on Livingstone’s task of exploring central Africa.

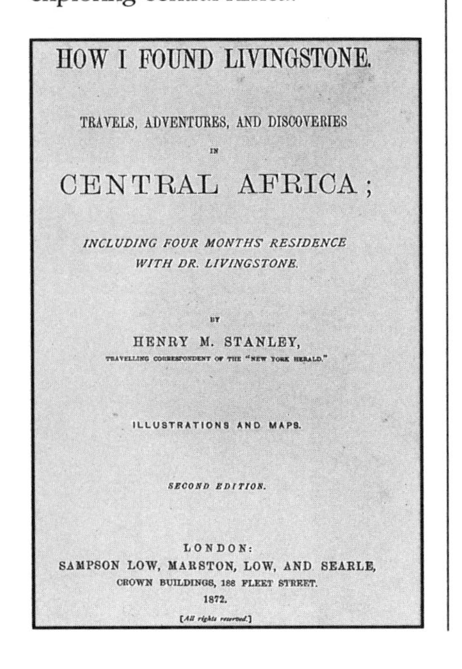

HOW I FOUND LIVINGSTONE.

TRAVELS, ADVENTURES, AND DISCOVERIES

IN

CENTRAL AFRICA;

INCLUDING FOUR MONTHS' RESIDENCE WITH DR. LIVINGSTONE.

BY

HENRY M. STANLEY,

TRAVELLING CORRESPONDENT OF THE “NEW YORK HERALD.”

ILLUSTRATIONS AND MAPS.

SECOND EDITION.

LONDON:

SAMPSON LOW, MARSTON, LOW, AND SEARLE,

CROWN BUILDINGS, 188 FLEET STREET.

1872.

[*All rights reserved.*]

In March 1873 Livingstone was ill and tired, pelted by downpours as he wandered around Lake Bangweulu. He was so weak that he had to be carried on a litter by Susi and Chuma. But he wrote defiantly in his diary, "Nothing earthly will make me give up my work in despair."

Eighteen months after his meeting with Stanley, Livingstone died near Lake Bangweulu, in present-day Zambia. One of the papers found among his effects is a sad testimony to the obsession of his final years. It is an announcement of his discovery of "the fountains of the Nile." The date and location are blank, waiting to be filled in.

Susi and Chuma, faithful to the end, preserved Livingstone's body and carried it 1,000 miles (1,600 kilometers) to the coast. It was shipped to London and buried in Westminster Abbey. Stanley attended his funeral. Shortly afterward, Stanley announced that he was mounting his own expedition to complete Livingstone's work of charting the lakes and rivers of central Africa. Stanley *had* been inspired by Livingstone's dedication and sense of high purpose, but he was also seduced by fame and and the glamour of African adventure.

In 1874 he arrived in Zanzibar at the head of one of the nineteenth century's largest and most ambitious expeditions. He and his 356 followers vanished into the interior and were not seen again for 999 days. In 1877 Stanley and 115 survivors emerged on the far side of the continent, at Boma on the mouth of the Congo River. Stanley's hero Livingstone had crossed Africa from west to east, and now Stanley had become the first European to cross it from east to west.

The trip had been a terrible one, full of slaughter, disease, hunger, and disappointment. Yet it made Stanley a hero to the many thousands of people who read newspaper accounts of his trip; he published his own lengthy version of the tale under the title *Through the Dark Continent*, and it was a best-seller. The workhouse boy from Wales was now recognized as the world's leading authority on the Congo region. At the request of King Leopold II of Belgium, who wanted to set up a trading company there, Stanley spent the years from 1879 to 1884 supervising the building of roads and trading forts in the Congo Basin. He was so strong and implacable that his work crews called him Bula Matari—"Rock Breaker."

Stanley made a fourth and final trip to Africa in 1887, this time to rescue another endangered European. Eduard Schnitzer, a German naturalist who had become a Muslim and adopted the name Emin Pasha, was holding a British fort near Lake Albert against the attacks of a Muslim army. The British government sent Stanley to relieve the siege against Emin Pasha. Stanley entered Africa by way of the Congo and struggled eastward for months through the dense rainforest, fighting with the native tribespeople for most of the way. By the time they reached Emin Pasha at Lake Albert, Stanley's troops were ragged and starving, able to offer little real help.

Worse than that, Stanley discovered that Emin Pasha had no intention of leaving his home. He felt perfectly safe; the panic in London over his well-being was unfounded. Stanley, however, would not take no for an answer. After all the pains he had taken to "rescue" Emin Pasha, he would look silly if he came back without the reluctant German. So he bullied—some accounts say he forced—Emin Pasha into withdrawing to Zanzibar with him. They arrived there in 1889.

That was Stanley's final African adventure. The book he wrote about it, *In Darkest Africa*, became another best-seller. Like Stanley's other books, *In Darkest Africa* helped create and promote an image that was widely held in Europe and the United States during the late nineteenth century: the intrepid white explorer who brought "enlightenment" to various "uncivilized" parts of the world, dauntlessly battling ignorance, savagery, and hostile Nature itself.

During his African journeys Stanley encountered dreadful hardships: floods, hunger, disease, and hostile Africans who had been stirred into a frenzy by the ravages of the slave trade. Yet he overcame all obstacles with the iron strength that earned him the nickname "Rock Breaker."

Stanley lived in the final days of European imperialism, or empire building. During the imperial era, France, Germany, the Netherlands, Portugal, and especially Britain built huge overseas empires by grabbing and holding colonies in Africa, Asia, and the Pacific. In the race for colonies, the nations of Europe divided all of Africa and most of southern Asia among themselves.

Colonies provided the home nations with raw materials, such as rubber, metal, wood, and cotton, that fed Europe's growing industries. The colonies also were convenient markets for industrial products, such as cloth, pottery, and machinery, that were made in Europe from those raw materials. The home nations sent administrators, traders, and missionaries to the colonies, which were in effect governed by these bureaucracies of white officials.

Most Europeans and Americans of the time regarded the native cultures of the different colonies—all of which had developed along quite different lines from European cultures—as fundamentally inferior to European cultures. Natives were often thought of as childish, savage, or even animallike. But although the Westerners exploited the colonies economically, many of them also felt a responsibility to "civilize" the natives—that is, to transplant their own languages, laws, and religions to the colonies.

This superior, imperialistic attitude was summed up by British poet Rudyard Kipling in his 1899 poem "The White Man's Burden." Although it was written about the United States and its colony in the Philippines, the poem expresses the views of many Europeans like Livingstone and Stanley, who saw the Africans in Kipling's terms: "new-caught, sullen peoples/Half devil and half child." According to Kipling, it was the white man's task to watch over and teach the native until the native could do so for himself—something that Kipling felt would take a long, long time. But although neither Stanley nor Kipling was aware of it when "The White Man's Burden" was written in the last year of the nineteenth century, colonialism and imperialism had almost run their course. Not many years later, colonies throughout Asia and Africa would demand and receive their independence.

As for Stanley, he died in 1904. He had retired from exploration to far greater comfort than Livingstone had ever known. He married, settled down in England, was elected to Parliament, and ended his days as a prosperous British gentleman.

CHAPTER 13

Spies on the Roof of the World

In the eighteenth and nineteenth centuries, India was Britain's largest and most important colony. As the British consolidated their hold over India, they sent army officers and soldiers, engineers, and government officials to administer the colony. Among the other tasks of colonial management, the British rulers of India carried out the most comprehensive mapmaking survey that had ever been done. Beginning in 1765, the Survey of India measured and mapped every part of the huge and crowded Indian subcontinent, the jewel in the British Empire's crown.

The survey began at the southern tip of India and moved painstakingly north. By 1834 the surveyors were within sight of the Himalayas, the immense mountain range that stretches for 1,500 miles (2,400 kilometers) along India's northern border. The British extended their survey right up to the foothills of the Himalayas, but the lands beyond remained frustratingly out of reach. They were unable to penetrate the wary, hidden kingdoms of Ladakh, Nepal, Sikkim, and Bhutan that nestled among the remote peaks of the Himalayas. The rulers of these ancient states, having seen India and southeastern Asia gobbled up by the European colonial powers, had resolved to keep Westerners out.

Most tantalizing of all these little-known lands was Tibet, high on a plateau behind the Himalayas, so lofty that it was called the Roof

An 1820 engraving showing the palace of a mountain prince presents a typically romanticized British view of the Himalayas—a wilderness of ice and rock on the northern border of India.

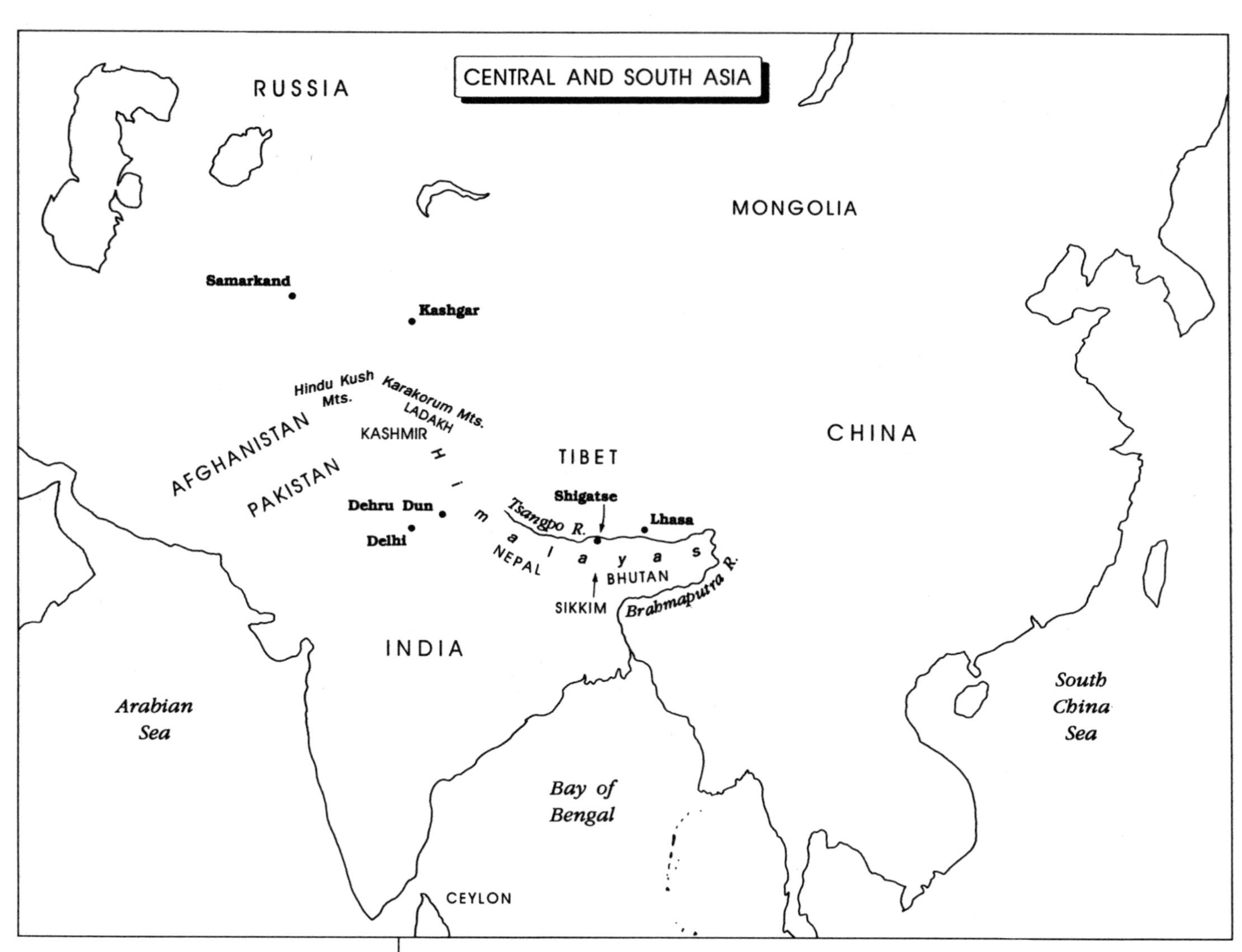

of the World—and so inhospitable to strangers that it was called the Forbidden Kingdom. A few bold or fortunate Westerners had managed to visit Tibet and the other almost inaccessible Himalayan kingdoms, but no British surveying party could hope to enter them. At best, surveyors would be unceremoniously turned back at the borders by armed guards; at worst, they would be executed.

Yet the British were determined to fill in the blank spaces north of India on their maps. They wanted to solve one of the great remaining geographical mysteries, and they disliked the untidiness and uncertainty of India's northern border. But they had a political reason as well. The Russian Empire was expanding rapidly through central Asia. Russian agents and scientists were working their way toward the Himalayas from the northwest, and the British wanted to prevent them from gaining control of lands along the Indian frontier.

Knowledge of these frontier lands was essential to British security, but how was it to be obtained? In the 1860s, Thomas G. Montgomerie, a captain in the Survey, came up with the answer. He suggested using natives from the Himalayan foothills and valleys to do undercover surveying. As he explained to the Asiatic Society of Bengal, a group of British scientists, "a European, even if disguised, attracts attention when travelling among Asiatics, and his presence, if detected, is nowadays often apt to lead to outrage." But natives of India, working under the control of the British government, could "travel freely without molestation in countries far beyond the British frontier."

For his first attempt on Tibet, Montgomerie selected two cousins, Nain Singh and Mani Singh, from the Indian Education Service. Nain Singh was a schoolteacher—a learned man or, in his native language, a *pandit.* The English pronounced this word "pundit" and applied the term to all the native surveyors. It has since entered the English language and means "a wise person" or "an expert."

Montgomerie spent two years teaching the Singhs how to carry out a topographical survey. They were to be disguised as Buddhists. But complete secrecy was vital. If they were caught measuring or mapping in the forbidden lands, they would certainly be killed.

Montgomerie devised a number of ingenious devices and tricks for the pundits. He taught them always to take 2,000 steps to the mile and to count their paces so that they could measure the distances they covered. They carried the prayer beads common among Buddhist pilgrims, but their beads were really aids for counting steps and performing mathematical calculations. The Singhs were equipped with portable Buddhist prayer wheels as hiding places for their notes and maps. Montgomerie taught them how to use watches, sextants, and thermometers; with these items the pundits could measure latitude and also altitude (the distance above sea level). But such instruments had to be kept carefully hidden in secret pockets in the surveyors' robes. Montgomerie prepared the Singh cousins to be geography spies on the Roof of the World.

In 1865 the Singhs left Dehra Dun, the Survey headquarters in northern India. They went north to the Hindu kingdom of Nepal. Mani Singh made his way to western Tibet, obtained data, and went back to Nepal. Nain Singh's journey was more hazardous. He was assigned to determine the exact location of Lhasa, the capital of Tibet.

Nain Singh attached himself to a caravan. Begging food from passersby, pretending to be a Buddhist monk on a pilgrimage, he reached Lhasa in January 1866. Upon his arrival he was interviewed by a panel of lamas, high-ranking Tibetan Buddhist priests, but they did not see through his disguise. Singh realized his danger, however, when he saw a Chinese trader beheaded for having entered Lhasa without permission.

He settled into an inn in Lhasa, and every night he crept onto the roof to make secret astronomical observations. Finally, fearing that some in the town had become suspicious of him, he left Lhasa with a caravan headed for Ladakh, in the western Himalayas. Two months later he sneaked away and crossed the mountains into Nepal.

Nain Singh reached Dehra Dun after a twenty-one-month absence. He had surveyed the 2,000-mile (3,200-kilometer) trade route between Nepal and Lhasa, he had measured the altitude of Lhasa to within 400 feet (120 meters) of the correct figure, and he had placed that mysterious capital accurately on the map at last.

The Survey of India next tried to solve another puzzle. A large river called the Tsangpo flowed through Tibet south of Lhasa, but no one knew where it came out. Nain Singh suggested that the Tsangpo was the same as India's Brahmaputra River, which flows out of the eastern Himalayas to empty into the Bay of Bengal in present-day Bangladesh. If the Tsangpo and the Brahmaputra were one and the same, the river must rise in Tibet, make a sweeping curve eastward across half the Tibetan plateau, and then flow through a mighty gorge somewhere in the Himalayas to reach the Indian flatlands. In 1880 the British surveyors embarked on an ambitious plan to find out whether the Tsangpo was also the Brahmaputra. They would send another secret agent into Tibet, and this agent would mark some logs and throw them into the Tsangpo—fifty logs a day for ten days. Surveyors would be stationed along the Brahmaputra in India to look for the logs.

Two men were selected for the mission: a Chinese lama and a pundit named Kinthup. A native of Sikkim in the eastern Himalayan foothills, Kinthup could neither read nor write, although he was well trained in the Survey's methods. He was to pretend to be the lama's servant. The two were supposed to sneak into Tibet, follow the Tsangpo for as much of its course as they could safely travel, and then launch the logs.

The lama proved perfidious. He prolonged the journey to Tibet by engaging in love affairs along the way, and he wasted time in Lhasa, eating and drinking with his fellow monks. During the journey east along the Tsangpo, he caused more trouble by flirting with the wives of villagers. He also began to treat Kinthup as though the pundit really were his servant—and Kinthup could not protest, for to do so would be to break his disguise. One day the lama simply sold Kinthup into slavery and disappeared.

Kinthup was put to work cutting grass for his new master's horses. He escaped after seven months and continued east along the Tsangpo, through increasingly jungly and difficult terrain. One day he was almost captured by his former "master." He sought sanctuary at a Buddhist monastery, and the head lama purchased him from his angry owner.

Four and a half months later, having received permission from the lama to make a pilgrimage, Kinthup hid in a nearby forest and laboriously cut and marked five hundred logs. He did not throw them into the river, however; first he had somehow to notify Captain H. J. Harman, his Survey boss, that the logs were coming. It was now eighteen months since Kinthup had left India, and Harman would have given up waiting for the logs by now. Kinthup needed to send a letter—but he could not write. So he hid the logs in a cave, worked in the monastery for two more months, and then asked for permission to make another pilgrimage. This time he went all the long, dangerous way back to Lhasa. It took him nearly a year.

Kinthup was lucky. In Lhasa he found an official from his native Sikkim who agreed to write a letter for him and smuggle it back to the Survey authorities. The letter said:

> Sir: The lama who was sent with me sold me to a Djongpen [a village headman] as a slave and himself fled away with the Government things that were in his charge. On account of which the journey proved a bad one; however, I, Kinthup, have prepared the 500 logs according to the order of Captain Harman, and am prepared to throw them 50 logs per day into the Tsangpo from Bipung in Pemake, from the fifth to the fifteenth day of the tenth Tibetan month of the year called Chhuluk, of the Tibetan calculation.

He did it, too, returning back along the eastward road to the secret cave to throw the logs into the river. Only then did he begin the long trek home. He arrived in September 1884.

Tragically, Kinthup's extraordinary devotion went unrecognized. The letter had gone astray, Captain Harman had died, and no one had been watching for his logs. His account of his adventures was

Kinthup was perhaps the most daring and persistent of all the Indian pundits who carried out undercover missions for the British Survey of India. His journey was so remarkable that for years people did not believe his story.

not even believed, and he left the Survey to become a tailor. It was many years before the mapmakers realized that Kinthup's story was true in every detail—and that the Tsangpo is indeed the Brahmaputra.

Kinthup was one of the last of the pundit explorers. Farmers, soldiers, schoolteachers—the pundits were men who in the ordinary course of Indian life might never have left their home villages. They became explorers by chance, plucked from obscurity by the hand of Montgomerie, who launched them on their extraordinary missions. For the most part, they returned to obscurity once those missions had been completed. Says Ian Cameron, author of *Mountains of the Gods: The Himalaya and the Mountains of Central Asia*:

> Kinthup was the archetypal Pundit, the epitome of those unselfish,unrewarded men—so anonymous that in many cases they were known only by their initials—who risked their lives for a cause they did not wholly understand and for a country other than their own. Even today and even in India and Pakistan their exploits are little known, yet in the history of exploration there have been few finer achievements.

But an indirect tribute to the pundits lives on in English literature. The hero of *Kim,* Rudyard Kipling's 1901 adventure tale, is a boy who is taught by Colonel Creighton of the British Secret Service to walk with measured paces and make secret observations. Creighton plans to send Kim into the Himalayas to spy on the Russians, in the geographical intrigue that Kipling called the Great Game. "So far as Kim could gather," Kipling wrote, "he was to be diligent and enter the survey of India as a chain-man" (a surveyor's assistant who measured distances with a length of chain).

Says Creighton to Kim, "Yes, and thou must learn how to make pictures of roads and mountains and rivers—to carry these pictures in thine eye till a suitable time comes to set them down on paper. Perhaps some day, when thou art a chain-man, I may say to thee when we are working together: 'Go across those hills and see what lies beyond.'" Those are the instructions that Montgomerie gave to his geography spies, the faithful pundits who were the first to map the mysterious kingdoms of the Himalayas.

PART FOUR

SURVIVORS AND ADVENTURERS

In the past, mapmakers sometimes extended the borders of the known world according to legend or conjecture: they drew in Prester John's kingdom, for instance, or filled the sea with finny, bewhiskered monsters. Similarly, the literature of travel and discovery contains many strange and colorful tales from the borderland where reality meets imagination. And even after the world began to be better known, and Prester John's kingdom had been banished from the map, people still hungered for tales of travel and adventure in places far from home.

CHAPTER 14

Great Escapes

Soldiers of the Ottoman Turks. The Ottomans seized young Hans Schiltberger and made him a well-traveled captive.

Some of the world's most enduring stories are tales of escape or survival. Readers in the 1970s and 1980s were captivated by *Alive,* the story of South American rugby players whose plane crashed in the Andes Mountains, and *Adrift*, a young man's account of seventy-six days lost at sea in an open boat. Fifteenth-century readers had the tale of Hans Schiltberger.

Hans (or Johann) Schiltberger was born in 1381 to a noble family in Bavaria, in southern Germany. At the age of thirteen he became the squire of a knight named Leinhart Richartinger, who followed King Sigismund of Hungary to war against the Ottoman Turks on the Hungarian frontier. Two years later, in 1396, the Ottoman sultan Bayezid I defeated the Christian forces at the battle of Nicopolis, and Hans was among the prisoners taken by the Turks.

Bayezid beheaded most of the prisoners, but Schiltberger and a few others were spared to become slaves; it is said that Schiltberger survived because the sultan's sixteen-year-old son took pity on him and pleaded for his life. Schiltberger was used as a runner to carry messages for the sultan. For six years he traveled with the sultan's retinue throughout the Ottoman territories in Egypt and the Middle East. He made at least one attempt to escape but was recaptured.

Schiltberger's fortunes changed again in 1402, when the Mongol warlord Timur Leng—known in the West as Tamerlane—overcame Sultan Bayezid in battle. Schiltberger became the property of Timur, who took him into the Mongol khanates, or kingdoms, of central Asia. He saw Armenia and the beginning of the great central Asian desert, although he went no farther east than the fabled city of Samarkand.

Timur could be shockingly brutal. When he captured the Persian city of Isfahan, he beheaded everyone over fourteen and built a tower with the heads. Then he burned the city. Schiltberger spent only three

Alexander Selkirk (opposite), marooned on a tiny Pacific island by his comrades, was the inspiration for the classic shipwreck story *Robinson Crusoe.*

William Dampier—pirate, explorer, and author—traveled three times around the world. He was marooned on an island in the Indian Ocean but escaped by making a dangerous sea voyage in a canoe.

years in his service, however. Timur died in 1405, and Schiltberger passed to a new owner, this time Timur's son Shah Rokh.

Chekre, a Mongol prince visiting Shah Rokh's court, was the next to acquire Schiltberger. He took him on a long journey through Siberia (Schiltberger may have been the first Western European to use this name for northeastern Russia) and then southern Russia. Finally, Schiltberger escaped and fled to Constantinople, the capital of Turkey. From there he was able to make his way home to Bavaria, where he arrived in 1427, after an absence of thirty-two years.

Schiltberger joined the court of a German duke and later published an account of his adventures. Although it contains much that is exaggerated or simply fabricated, such as his tale of giant warrior women in Tartary, Schiltberger's book also contained a great deal of reliable new information. He was the first European, for example, to identify the holy Islamic city of Medina in Arabia as the burial place of the prophet Muhammad. He provided European kings and merchants with details about the military organization and commerce of the Ottoman Empire and the Mongol khanates. Many readers, though, enjoyed his story simply as a rousing adventure tale.

Another such tale—and one of the most commercially successful books of all time—was Daniel Defoe's *Robinson Crusoe*, published in 1719. The story of a shipwrecked mariner cast ashore on a deserted isle, *Crusoe* was fiction, but its author drew heavily from the real-life experiences of two castaways, William Dampier and Alexander Selkirk. Unlike Crusoe, neither was washed ashore after a wreck. Instead, each was marooned, or put ashore, by his companions.

Dampier was an English pirate-turned-explorer who made a voyage of discovery to Australia for the British Admiralty and then turned again to piracy. He was remarkably well traveled; between 1680 and his death in 1715 he made three trips around the world. On the first of these voyages, he and three shipmates who hated and feared their captain asked to be marooned on a small island in the Indian Ocean. They obtained a canoe from the natives and sailed across 200 miles (320 kilometers) of stormy sea to Sumatra.

In 1697, after returning to England, Dampier published his journal as *A New Voyage Round the World*. The book contained an account of Dampier's marooning in the Indian Ocean; it also told of how his ship had picked up a Central American Indian who had been marooned for three years with nothing but a knife and a gun on the small island of Más a Tierra in the Juan Fernández Islands, off the coast of Chile. Dampier's book came to the attention of Daniel Defoe and helped give him the idea for *Robinson Crusoe.*

Another inspiration for *Crusoe* was the story of Alexander Selkirk, who was also involved with Dampier and Más a Tierra. Selkirk was a touchy, proud Scottish sailor who in 1704 served on the *Cinque Ports,* a vessel that was making a circumnavigation of the globe; Dampier was captain of another ship in the expedition. While the *Cinque Ports* was anchored in the Juan Fernández Islands, Selkirk and his captain quarreled, and Selkirk cried out that for all he cared the ship could go on without him. The captain seized upon these angry words and decided that he *would* proceed without Selkirk. The Scotsman was put ashore with only his sea chest.

Marooning was a dreadful fate. At worst, the marooned sailor might starve, die of thirst or disease, or be killed by hostile natives; at best, he faced isolation, perhaps for the rest of his life. Many marooned men went mad from loneliness.

Selkirk was fortunate. Like the marooned Indian before him, he survived. Más a Tierra had water, food (edible plants and lots of turtles and crabs), and no inhabitants or dangerous beasts. There were wild goats and wild cats, and Selkirk eventually captured and tamed some of each—the goats for food, the cats for companionship. He built

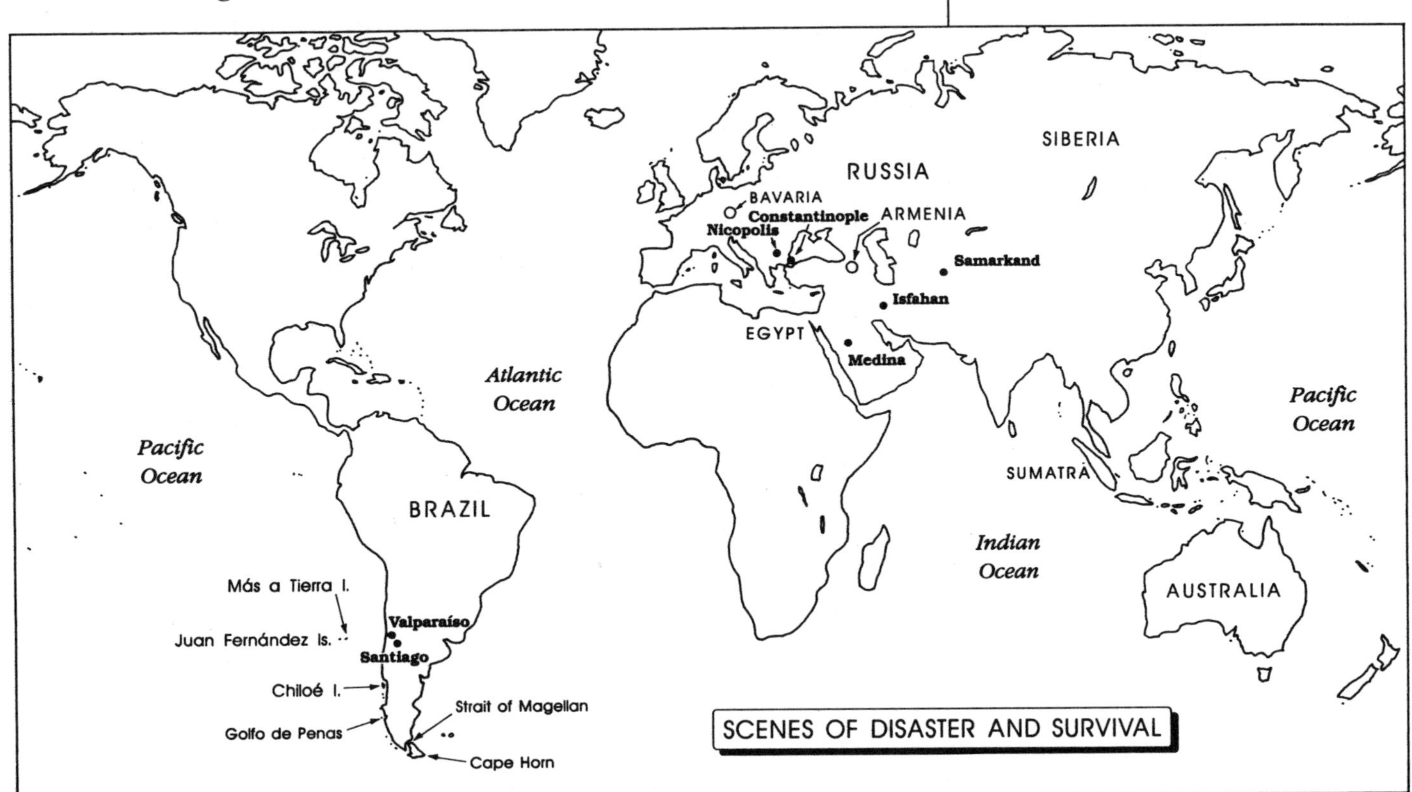

a hut, sewed new clothes for himself out of goatskins, and read the Bible; in short, he lived a life very similar to that described in *Crusoe,* except that he found no loyal native sidekick. He was rescued after four years by a British expedition under the command of Woodes Rogers, with Dampier as one of the captains. Selkirk's story was told in Rogers's 1712 book *A Cruising Voyage round the World,* which Defoe read. Más a Tierra now belongs to Chile and is sometimes called Isla Robinson Crusoe.

Selkirk's adventure inspired Defoe, whose novel in turn inspired many successors, including Robert Louis Stevenson's *Treasure Island* and Johann David Wyss's *The Swiss Family Robinson.* But the adventures of shipwreck victim John Byron have never inspired a popular novel: perhaps they are too gruesome.

In 1741 the eighteen-year-old Byron was a midshipman on the British Navy ship *Wager,* part of a squadron that was to sail around Cape Horn at the tip of South America to attack Spanish settlements on the continent's west coast. As the *Wager* was rounding the Horn, a gale blew the ship out into the Pacific Ocean and tore away its sails. The ship became separated from the rest of the fleet and limped northward along the coast through foul weather.

When the skies cleared, the men saw that the *Wager* had gotten inside a large bay, although no such bay appeared on their charts. Glaciers, blank cliffs, and snow peaks were ranged around them; on the shore a line of white marked the breaking of fierce waves. There was no sign of shelter or habitation. They tried desperately to get out of this grim trap, but bad winds, illness among the crew, and a serious injury to the captain kept the ship from leaving the bay.

At dawn on May 14 the ship was driven aground on jagged rocks. The men on deck began screaming, weeping, and praying; those below deck drowned instantly when the hold was flooded with bitterly cold water. Byron and 139 others made it safely to shore, but some crewmen remained on the rockbound wreck and refused to share the ship's provisions. Already the shipwreck victims were divided into hostile camps—and the hostility was to grow.

The months that followed were a violent drama of mutinies, betrayals, murders, arrests, court-martials, fights with the local Indians (who then disappeared from the region), and repeated but unsuccessful attempts to reach civilization. Rations were scarce, and the unforgiving terrain offered scanty forage; everyone was constantly on the edge of starvation.

The men were in a particularly bleak place, far from any source of aid. They called the bay the Gulf of Afflictions, and it bears that name today (Golfo de Peñas in Spanish). It is located between 47° and 48° S off the coast of Chile. The *Wager* had run aground on a small island, which the victims called Wager Island; it is now called Isla Byron. The region is still almost untouched. Although it possesses an austere beauty, it is a cold, rainy, and dismal place.

In the end, the survivors of the wreck—their numbers steadily diminishing—split into two rival groups. The mutineers managed to take the ship's longboat out of the bay, along the treacherous coast, through the Strait of Magellan, and up the east coast of South America to the Portuguese settlement in Brazil. The other group, led for the most part by Byron because the captain was seriously ill, obtained the help of some Indians and took a barge up the coast to the island of Chiloé, where there were Spanish officials.

With the wreck of the *Wager* in the background, the survivors take stock of the dismal coast where they have been cast ashore.

Spain and Britain were enemies, and the Spanish were well aware that the *Wager* had come to South America to attack them. Byron and the two remaining survivors were arrested, but they almost felt it was worth it to be warm, clothed, and fed after a year of misery. The prisoners were well treated, and eventually they were sent north to Valparaíso and then to Santiago. There they languished for two years, befriended by the local Spanish aristocracy, while the governments of Spain and Britain arranged an exchange of prisoners. Finally, on December 20, 1744, they sailed from Valparaíso for home.

Byron reached England in 1745 and was properly regarded as the hero of the whole affair. He had shown himself to be loyal, capable, and tough. In addition, his account of his travails provided the British with considerable information about a dangerous region that had not previously been visited. Based on his report, the Admiralty was able to complete its charts of the South American coastline.

Byron went on to enjoy a long and successful naval career, during which his fortitude in the face of gales caused his men to nickname him "Foul-Weather Jack." Perhaps he felt that, having survived the wreck of the *Wager,* he had nothing more to fear from storms at sea. Byron's only contribution to literature is an indirect one: the many references to shipwrecks that occur in the epic poem *Don Juan,* written by his grandson, George Gordon, Lord Byron.

CHAPTER 15

Swashbuckling

Centuries ago the word *swashbuckler* referred to someone who, in battle or in a fight, struck his shield (sometimes called a buckler) with his sword to make a loud noise that might frighten his opponent. The term gradually came to be used for any loud, boisterous ruffian. Today it often means simply "a colorful adventurer," but the earlier meaning of "a bragging roughneck" applies very well to Arnold Henry Savage Landor, who visited nearly every interesting place in the world—and made himself out to be a hero in every one of them.

Savage Landor, as he liked to be called, was born in 1867, the grandson of the poet Walter Savage Landor and the son of wealthy parents. He grew up in Italy and was, by his own account, a prodigious student, athlete, and mountaineer by age fifteen. He read a book about the Nile explorations and dreamed of having adventures. "Little did I think at that time that in the way of adventure I should have in later years enough to fill the lives of twenty men," he wrote.

Modesty was not one of Savage Landor's virtues. He reported that when he took up a paintbrush in Cornwall, he revealed himself as an artistic genius. He was both proud and stubborn. For example, he refused to wear rugged clothing or use special equipment when he traveled or climbed—no boots, parkas, climbing ropes, or sun hats. He felt that a true British gentleman should be able to impose his way of doing things on the rest of the world, not the other way around.

He visited Japan from 1888 to 1889 and went to the island of Hokkaido, home of Japan's original inhabitants, the Ainu people. His book about the trip was well received, and in 1896 he set off for Tibet, which was still regarded by nearly everyone as the most remote and inaccessible of all destinations.

Boastful and snobbish, Savage Landor (opposite) was a roving adventurer who made himself famous with dramatic stories of his exploits in remote places. Some of these stories were actually true.

Barefoot and bound with ropes, Landor is dragged by his Tibetan captors. When he published the tale of his adventures in Tibet, he indignantly listed all the injuries he had suffered.

In a Forbidden Land, the story of his travels in Tibet, was published in 1898. Its startling subtitle was *An Account of a Journey in Tibet, Capture by the Tibetan Authorities, Imprisonment, Torture, and Ultimate Release*, and its contents more than lived up to that grandiose billing. According to Landor, he had been dragged behind a pony by his neck, burned with hot irons, and stretched on the rack. He had then escaped over the Himalayas, been hunted by a huge army, fallen down a mountain, forded ice-choked torrents, eaten nettles, discovered the true sources of three major rivers (perhaps he remembered dreaming over his Nile book as a youngster), drawn superb new maps of Tibet, and to top it all off had climbed a mountain to the unheard-of height of 22,000 feet (6,666 meters) without climbing boots, an ice ax, or a rope. Not only that, but he had made the climb in record-breaking time with a heavy bag of silver coins on his back. What a man!

But his most satisfying accomplishment, he wrote smugly, was "teaching the non-Asiatic peoples to pronounce correctly the name of the greatest mountain range on earth, viz Himahlya and not Himalaya, a meaningless distortion of a poetic word." The book included a list of every scratch and bruise inflicted on his rugged person by the fiendish Tibetan torturers.

The specialists of Britain's Royal Geographical Society and the *Alpine Journal* were highly critical of Savage Landor's wild claims. The geographers pointed out that his "new maps" were nothing more than fancy versions of the maps the society had given him before his departure. The mountain climbers pointed out that it would have been utterly impossible for Landor or anyone else to cover the distance he claimed to have covered on his mountain within the stated time. In addition, some people felt that his arrogant, brutal, chauvinistic attitude made all British travelers look bad.

But the public loved him. The book was a best-seller and Savage Landor was the most sought-after lecturer in Europe and America. He did not wait long, however, before starting his next expedition; in fact, he became a sort of perpetual traveler. He returned to the Himalayas—or Himahlya, as he would have said—in 1899 and again claimed to have set a climbing record, although the mountain that he said was 23,000 feet (6,970 meters) high was later shown to be only 16,500 feet (5,000 meters). He then blustered into Tibet, although a five-thousand-man army tried in vain to stop him, and gave "a sound whipping" to those who had earlier mistreated him.

Later journeys carried him through the deserts of Iran and the jungles of the Philippine Islands, across Africa ("at its widest point," he boasted), and across South America, where he hired criminals and

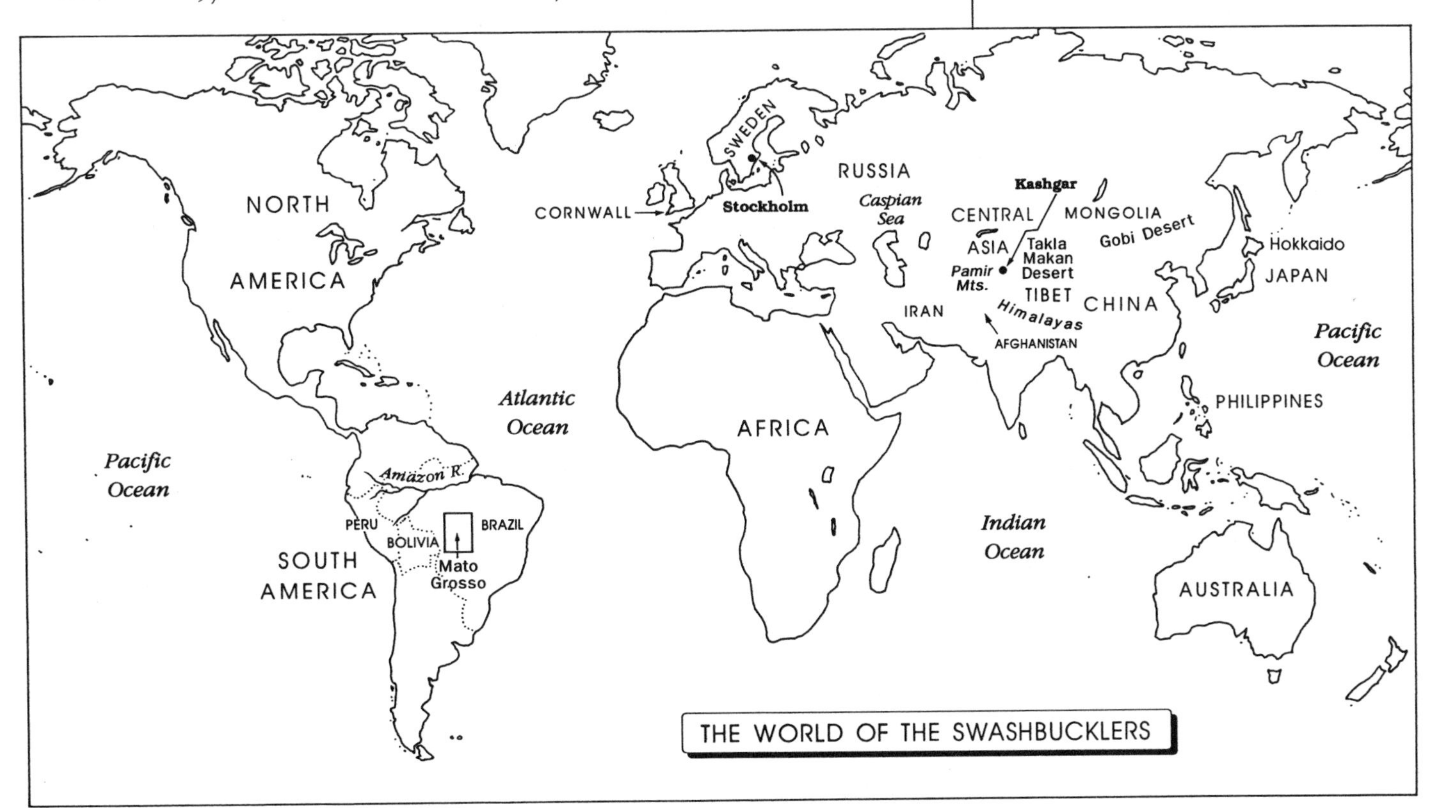

headhunters to serve as his porters and then masterfully quelled their mutinies in a series of bloodcurdling adventures. The South American trip was littered with incident: the finding of an enormous fossilized dinosaur (which later travelers were never able to find), a canoe trip down a river filled with "Stupendous Vortexes" and "Awe-inspiring Whirlpools," an attack by flesh-eating ants. It was Savage Landor's last great adventure. He was no longer young, and he had come too close to death to shrug off the danger. He spent the rest of his life traveling around Europe, visiting royalty and the Pope. He died in 1924, shortly after finishing a biography that travel historian John Keay calls "surely one of the most outrageous books ever published."

Savage Landor was not an outright liar—at least, most of the time he was not. He was, however, addicted to exaggeration. There was truth at the core of all his larger-than-life stories. For example, he really was imprisoned, and to some extent abused, by the Tibetan authorities, and he did have remarkable adventures escaping from them across the Himalayas. But Savage Landor could not resist the impulse to stretch the truth as far as it would go, and sometimes he stretched it beyond the breaking point. He was the swashbuckler supreme.

He was also one of a new breed of accidental explorers. These were people who traveled not to open up new trade routes or to extend geographical knowledge—on business, as it were—but rather for the sheer excitement and adventure of it. They often recounted their exploits in books and lectures. The more mysterious, dangerous, and remote a destination, the better these adventurers—and their readers—liked it. Therefore, in the process of circling the globe and accumulating travel tales, the adventurers often slipped into territory that was all but unknown and highly hazardous. Some of them became what might be called accident-*prone* explorers, reeling or dashing from one mishap or hairbreadth escape to the next, all the while taking notes for the armchair explorers at home.

At the same time that Savage Landor was swaggering through the world's remote places and giving boastful accounts of his adventures, another European adventurer was exploring many of the same places. His stories were as thrilling as Savage Landor's, although they were somewhat more sober and honest.

His name was Sven Hedin. Born in Stockholm, Sweden, in 1865, Hedin grew up fascinated by the idea of travel and adventure. At the

Sven Hedin began as a swashbuckler but became one of the world's foremost experts on the geography and history of central Asia. He was a proud, withdrawn man who prided himself on enduring rugged journeys and appalling hardships.

age of twenty he accepted a job as tutor to the children of a Swedish family who lived in Russia, on the shore of the Caspian Sea, just so that he could visit that part of the world.

His trip to Russian Turkestan, as the region around the Caspian Sea was called, was the beginning of a long passion. For the rest of his life, Hedin was enthralled by the deserts, mountains, and steppes of central Asia—the lands of the old Silk Road and the Mongol Empire, today divided among China, Mongolia, and the Russian commonwealth. He spent many years traveling in this part of the world, and he seemed to thrive on the isolation and difficulty that his travels imposed upon him. He once wrote, "Those who imagine that such a journey in vast solitude and desolation is tedious and trying are mistaken. No spectacle can be more sublime. Every day's march, every league brings discoveries of unimagined beauty."

Hedin did much to capture the beauty of the remote places he visited and bring it back to the rest of the world. He took many black-and-white photographs and made detailed, highly accurate maps. He also made hundreds of sketches and paintings that provide a vivid, colorful record of the lives of villagers, nomads, monks, and caravan masters in the final years before the twentieth-century civilization of the outside world reached and changed them.

Hedin's travels nearly ended prematurely. In 1895 he and his servant Kasim formed a small camel caravan to cross the Takla Makan Desert in what is now western China. The Takla Makan is a treacherous desert, where wind endlessly shifts the rolling sand dunes, sometimes hiding the road and burying the few wells and waterholes. Hedin planned to cross it at a point where it had never before been crossed, solely on the strength of a vague rumor that water could be found on the other side.

They set out from the oasis city of Kashgar. About 200 miles (320 kilometers) to the east, they began to run low on water. Soon all the water was gone. The camels began to die.

The men tried to drink stove fuel and sheep's blood, but could not stomach either liquid. One by one, they fell by the wayside, until only Hedin and Kasim were left to stagger on, "like sleepwalkers," Hedin later wrote, across the dunes. Then, to their despair, they discovered that they had been walking in circles. But just when hope seemed lost, the new light of sunrise showed them a dark green line on the distant horizon: a row of trees that could only mark the course

During one of his later journeys, Hedin sketches a Mongolian woman.

of a river. Kasim collapsed, but Hedin pushed himself onward. When he became too weak to walk, he crawled. Finally he reached the river.

It was dry. The riverbed consisted only of cracked brown mud. He kept crawling along the riverbed, hoping for a trickle of water, until ahead of him he heard a bird splashing, and the welcome sound led him to a pool. He drank and then filled his boots with water for Kasim. They survived, and Hedin later crossed the Takla Makan again, only to be plagued by snowstorms and sandstorms.

Hedin never turned back in the face of danger. He once said of fellow explorer Aurel Stein, who had decided to retreat with his own caravan from a desert crossing when the going became rough, "In a similar situation I should never have made such a decision. I should have continued through the desert. It might have been the death of me and my men...but the adventure, the conquest of an unknown country, and the struggle against the impossible, have a fascination which draws me with irresistible force." Hedin felt that facing danger gave him a kind of nobility. He "galloped headlong against the physical challenge of unknown Asia as though he were entering the lists of a tournament," says historian Timothy Severin in *The Oriental Adventure*.

The Takla Makan episode was not the only close call of Hedin's career. Traveling in the Pamir Mountains of Afghanistan, he decided to climb a tall, steep mountain that the local people called Father of Ice. His native servants protested that the climb was too dangerous. It was the middle of avalanche season, and Hedin had no climbing equipment. But he tried repeatedly to scale the summit, driven back every time by heavy snowfalls that almost killed him. Finally the relentless glare of light on the icy slopes afflicted him with temporary snow blindness. His servants—who might have been tempted to abandon him after he had risked their lives—patiently led him by the hand back to his camp.

These escapades made exciting reading, but Hedin grew tired of being dismissed as a show-off and an adventurer in the Savage Landor style. He wanted to be taken seriously as an explorer. He embarked on extensive journeys in the barren plateau country of northern Tibet and the Himalayan ranges along Tibet's southern border, recording his travels in long, scholarly volumes about places no European had yet seen. Following the example of the pundits in the Indian Survey, he entered Tibet disguised as a monk with a Buddhist rosary and an image of the Buddha. On his first attempt to enter Tibet he was detected and turned back by the Tibetan authorities, but eventually he succeeded.

In addition to his geographical surveys, Hedin made some important archaeological discoveries. Led by local guides, he unearthed the sites of several buried towns, and he recovered documents dating from the third century A.D. He deeply felt the mysterious thrill of such discoveries. After finding the first of the buried towns, he wrote, "Here I stand, like the prince in the enchanted wood, having wakened to new life the city which has slumbered for a thousand years."

Hedin's achievements were recognized and honored by the learned societies that scoffed at Savage Landor. The Royal Geographical Society of London awarded him a gold medal (but later took it back because he sided with Germany in World War I). He was made a knight in Britain and a nobleman in Sweden. When he was in his sixties, he made two further expeditions into central Asia, surveying western China and the Silk Road caravan route. The first of these, which began in 1928, was the largest exploring expedition that had ever been undertaken. With a corps of Swedish and Chinese scientists and assistants, Hedin set up camps across Asia from the Caspian Sea to the Great Wall of China. In six years they produced fifty-six books and thousands of pages of notes on the plants, animals, fossils, geology, weather, languages, history, and geography of central Asia.

A Tibetan musician drawn by Hedin in 1907.

Sven Hedin died in 1952 in Sweden. He had started swashbuckling to satisfy his obsession with travel and adventure, but in the end he became a true explorer.

Savage Landor demanded attention and admiration; Hedin appreciated it. Twentieth-century swashbuckler Percy Fawcett received more attention than either of them, but his attitude about it is a mystery, as is his fate.

Fawcett was an English army officer who in 1906 was sent to South America to help survey the borders of Bolivia, Peru, and Brazil. He became so passionately fond of exploring and surveying that he quit the army in 1910 because it refused to let him continue doing what he loved. He worked for the Bolivian government as a civilian surveyor until 1914, when he left the survey and rejoined the army to fight in World War I.

He was back in South America in 1920. By this time he had developed the theory that the still-unexplored interior of central and western Brazil, a vast region called the Mato Grosso, contained a lost race or lost civilization. The theory was based on his own observations and also on the accounts of previous travelers who reported seeing strange people or hidden cities in the jungle. The first Europeans to travel the Amazon, Francisco de Orellana's crew, had in fact reported meeting "white Indians" somewhere along the river in 1542. This and other rumors and legends—the modern version of the El Dorado story—fueled Fawcett's curiosity.

The Brazilian government offered to sponsor an expedition to find the lost city of the Mato Grosso. Fawcett began the investigation but was unable to complete it because the exhaustion of his companions forced him to turn back. He vowed that for his next trip he would have comrades he could count on.

His choice of comrades fell on his twenty-one-year-old son Jack and Jack's friend Raleigh Rimmell. On April 20, 1925, the three men vanished into the Mato Grosso. Their course was to take them north through the Mato Grosso and then east across the widest part of the continent to the Atlantic shore. They were never seen again. One letter came out of the jungle, carried by an Indian. It was dated May 29. From that time on, the jungle was silent.

By 1928 Fawcett's friends and backers were worried that he might need rescue. The first of many rescue expeditions followed Fawcett's path to his last known camp; its commander believed that Fawcett and the others had been killed by Indians, although there was no firm evidence. A number of other rescue attempts were made, and for a time "the search for Colonel Fawcett" was almost as much a topic of public concern as the searches for Franklin and Livingstone had been.

Legends of a lost city in the jungle drew Percy Fawcett and his doomed party into the heart of Brazil's Mato Grosso.

Rumors of Fawcett's status and whereabouts flew around South America for years. In 1932 a Swiss trapper came out of the jungle with a story of having met a European man who was the prisoner of an Indian tribe. He identified the captive as Fawcett. Several years later two rumors surfaced: that three white men were being held captive deep in the jungle, and that three white men had been murdered. Finally it appeared that the mystery was solved in 1950, when the chief of an Indian tribe near Fawcett's last camp was dying. He confessed to having killed Fawcett. Bones from "Fawcett's grave" were turned over to the Brazilian government with great ceremony and enshrined in a museum in Rio de Janeiro. Later studies proved, however, that the bones were not Fawcett's and probably belonged to an Indian. So the last word on Fawcett's fate has not been written—and probably never will be.

But the *best* words on Fawcett's fate were written by Peter Fleming, brother of the Ian Fleming who created James Bond. In 1932 Peter Fleming answered a newspaper ad seeking volunteers for a "Rescue Colonel Fawcett" expedition. The expedition failed to turn up the missing colonel, but it did result in one of the funniest travel books ever written, Fleming's *Brazilian Adventure,* published in 1933. Letting the reader know that this travel book takes a somewhat ironic approach to its subject, Fleming's introduction sticks a sly pin into the pompous windbaggery of swashbucklers like Savage Landor:

> Differing as it does from most books about expeditions, this book also differs from most books about the interior of Brazil. It differs in being throughout strictly truthful. I had meant, when I started, to pile on the agony a good deal; I felt it would be expected of me. In treating of the Great Unknown one has a free hand, and my few predecessors in this particular field had made great play with the Terrors of the Jungle. The alligators, the snakes, the man-eating fish, the lurking savages, those dreadful insects—all the paraphernalia of tropical mumbo jumbo lay ready to my hand. But when the time came I found that I had not the face to make the most of them. So the reader must forgive me if my picture of the Mato Grosso does not tally with his lurid preconceptions.

AFTERWORD

The *Why* of Exploration

People have explored this world for many reasons. Some, like the Chinese Buddhist pilgrims and the Christian missionaries, have had religious motives. Some, like Chang Ch'ien, or Pero da Covilhã, or Captain James Cook, or Sir John Franklin, or the pundit Kinthup, were simply following orders. Often the motive for those orders was political or economic. Great exertions are possible when the pride of nations, or their income, is at stake. Financial motives are always strong; something as simple as the desire for cheaper pepper was behind a centuries-long period of exploration and expansion that propelled Europeans into Africa, Asia, and the Americas. And there are always individuals—the Pizarros, the Martin Frobishers, and the Marco Polos—who will venture upon great risks for the chance of great gain.

Some explorers have been driven by pride and curiosity. Erik the Red and his son Leif Eriksson come to mind, as do Henry Morton Stanley, Savage Landor, and Percy Fawcett. And many others have been made explorers or adventurers by circumstance, like Naddod the Viking, Francisco de Orellana, and Hans Schiltberger. But except for the victims of sheerest accident, most explorers and discoverers have probably had very mixed motives, compounded of greed, duty, patriotism, pride, and curiosity.

The year 1992 marks the five hundredth anniversary of the voyage of Christopher Columbus, the most accidental explorer of all. It is no longer possible to view Columbus and the other figures of the great age of exploration with undiluted admiration, as earlier generations did. There is growing recognition that what the Elizabethans and Victorians considered to be heroic exploration may also be viewed as exploitation and arrogance. And many of the heroes of discovery have been shown by recent studies and biographies to have been flawed—driven into the distant and lonely places of the world, perhaps, by the very qualities that made them unhappy or unsuccessful at ordinary life.

Yet no matter how much the motives and actions of explorers are dissected and reexamined over the years, there is one aspect of the history of discovery that can never be completely discounted. As basic and intangible as human nature, it is the love of adventure, the impulse to go to extremes, the need to know what is unknown. That spark has burned in many explorers, accidental and otherwise. In his 1898 poem "The Explorer," Rudyard Kipling characterized it as the voice that endlessly whispers:

> "Something hidden. Go and find it. Go and look
> behind the Ranges—
> Something lost behind the Ranges. Lost and
> waiting for you. Go!"

Chronology

138 B.C.
Chang Ch'ien begins 13 years of travel in central Asia.

2nd century A.D.
Ptolemy writes his *Geography*.

around 400
Fahsien begins a 15-year pilgrimage through South Asia.

629
Hsüan-tsang begins a 16-year trip through central Asia and India.

around 800
Vikings reach the Faeroe Islands.

mid-9th century
Several Vikings sight Iceland.

900 or 930
Gunnbjorn Ulfsson sights Greenland.

982
Erik the Red explores Greenland.

986
Bjarni Herjolfsson sails along the North American coast.

around 1001
Leif Eriksson explores North America and winters in Vinland.

1145
Bishop Otto of Germany writes about Prester John, a legendary king thought to rule somewhere in Asia.

13th and 14th centuries
Marco Polo and other European travelers cross Asia.

1410
The first Latin translation of Ptolemy's *Geography* appears in Europe.

1483
Portuguese navigator Diogo Cão reaches the Congo River mouth.

1492-93
Columbus makes his first voyage to America, which he thinks is Asia.

1493
Portuguese spy Pero da Covilhã travels to Abyssinia.

1493-96
Columbus makes a second voyage to America and starts a Spanish colony on the island of Hispaniola.

1497
John Cabot, an Italian navigator working for Britain, explores the coast of Newfoundland Island.

1498
Columbus makes a third voyage and explores the coast of South America; Cabot is lost at sea.

1500
On his way from Portugal to India, Pedro Álvars Cabral lands in Brazil.

1501 and 1503
Amerigo Vespucci explores the coast of Brazil.

1502-04
Columbus makes a fourth voyage and explores part of the coast of Central America.

1506
Columbus dies in Spain, still convinced that he had reached Asia.

1507
German mapmaker Martin Waldseemüller names America for Amerigo Vespucci.

1513
Juan Ponce de León lands in Florida.

1519-22
A Spanish fleet under Ferdinand Magellan makes the first trip around the world.

1540-42
Francisco Vásquez Coronado explores the American Southwest; his men are the first Europeans to see the Grand Canyon.

1541-42
Francisco de Orellana leads the first European voyage down the length of the Amazon River.

1576-78
Englishman Martin Frobisher makes three voyages across the North Atlantic to Baffin Island.

1595 and 1617
British adventurer Sir Walter Raleigh searches for El Dorado.

1616
Dutch merchant Dirck Hartog is the first European to visit Australia.

1642
Dutch explorer Abel Tasman circumnavigates Australia, landing in Tasmania and New Zealand.

1697
William Dampier publishes *A New Voyage Round the World.*

1704-08
Alexander Selkirk is marooned on an island off the coast of Chile.

1712
Woodes Rogers publishes *A Cruising Voyage Around the World.*

1719
Daniel Defoe publishes *Robinson Crusoe.*

1741
John Byron and others are shipwrecked in Chile; Byron returns to England in 1745.

1768-79
James Cook of the British navy makes three voyages to the Pacific; he is the European discoverer of Hawaii and the first navigator to cross the Antarctic Circle.

1826-27
Jedediah Smith and fellow beaver trappers are the first white men to cross the Mojave Desert to California and the first to cross the Great Nevada Basin.

1845
British explorer Sir John Franklin sets out to discover the Northwest Passage; he dies in the Canadian Arctic in 1847.

1853-56
Scottish missionary-explorer David Livingstone crosses Africa from west to east and finds the Victoria Falls on the Zambezi River.

1859
The first traces of Franklin's missing expedition are found.

1865-66
Pundit Nain Singh explores Lhasa, the capital of Tibet, on a mapping mission for the British.

1866
Livingstone sets off to find the source of the Nile River.

1871
Reporter Henry Morton Stanley meets Livingstone in Tanzania.

1873
Livingstone dies in Zambia.

1874-77
Stanley crosses Africa from east to west.

1880-84
Pundit Kinthup explores the upper Brahmaputra River in Tibet.

1893-1930s
Swedish explorer Sven Hedin explores Mongolia, Tibet, and central Asia.

1907
Hungarian explorer Aurel Stein discovers centuries-old documents in a cave in western China.

1925
Percy Fawcett and companions disappear in South America while searching for a lost civilization.

Further Reading

ABOUT EXPLORATION IN GENERAL

Boorstin, Daniel J. *The Discoverers.* New York: Random House, 1983.

Charnan, Simon. *Explorers of the Ancient World.* Chicago: Childrens Press, 1990.

Connell, Evan S. *The White Lantern.* New York: Holt, Rinehart and Winston, 1980.

Day, Alan E. *Discovery & Exploration: A Reference Handbook.* London: Bingley, 1990.

Divine, David. *The Opening of the World: The Great Age of Maritime Exploration.* New York: Putnam, 1973.

Flaum, Eric. *Discovery: Exploration Through the Centuries.* New York: Smithmark, 1991.

Franck, Irene, and David Brownstone. *To the Ends of the Earth: The Great Travel and Trade Routes of Human History.* New York: Facts on File, 1984.

Herrmann, Paul. *Conquest by Man.* New York: Harper, 1954.

———. *The Great Age of Discovery.* New York: Harper, 1958.

Humble, Richard. *The Explorers: The Seafarers.* Alexandria, Va.: Time-Life Books, 1978.

Jackson, Donald Dale. "Who the Heck Did 'Discover' the New World?" *Smithsonian,* Sept. 1991, 76-85.

Keay, John. *Eccentric Travellers.* London: John Murray, 1982.

———. *Explorers Extraordinary.* Los Angeles: Tarcher, 1986.

Leslie, Edward E. *Desperate Journeys, Abandoned Souls: True Stories of Castaways and Other Survivors.* Boston: Houghton Mifflin, 1988.

Lewis, Richard S. *From Vinland to Mars: A Thousand Years of Exploration.* New York: New York Times Book Co., 1976.

Lomask, Milton. *Great Lives: Exploration.* New York: Scribners, 1988.

Matthews, Rupert. *Explorer.* New York: Knopf, 1991.

Maxtone-Graham, John. *Safe Return Doubtful: The Heroic Age of Polar Exploration.* New York: Scribners, 1988.

Moorehead, Alan. *The Blue Nile.* New York: Harper & Row, 1962.

———. *The White Nile.* New York: Harper & Row, 1960.

Morison, Samuel Eliot. *The Great Explorers: The European Discovery of America.* New York: Oxford University Press, 1972.

Newby, Eric. *The World Atlas of Exploration.* New York: Crescent Books, 1975.

Newby, Eric, ed. *A Book of Travellers' Tales.* New York: Penguin, 1985.

Parker, John. *Discovery: Developing Views of the Earth from Ancient Times to the Voyages of Captain Cook.* New York: Scribners, 1972.

Pennington, Piers. *The Great Explorers.* London: Bloomsbury Books, 1979.

Reader's Digest Books. *Great Adventures That Changed Our World: The World's Great Explorers, Their Triumphs and Tragedies.* Pleasantville, N.Y.: Reader's Digest Association, 1978.

Robinson, Jane. *Wayward Women: A Guide to Women Travelers.* New York: Oxford University Press, 1990.

Ryan, Peter. *Explorers and Mapmakers.* London: Lodestar, 1990.

Sandak, Cass. *Explorers and Discovery.* New York: Franklin Watts, 1983.

Scammel, G.V. *The First Imperial Age: European Overseas Expansion, 1400-1715.* London: Unwin Hyman, 1989.

Severin, Timothy. *The Oriental Adventure: Explorers of the East.* Boston: Little, Brown, 1976.

Stefansson, Vilhjalmur. *Adventures in Error.* New York: McBride, 1970.

Stefansson, Vilhjalmur, ed. *Great Adventures and Explorations: From Earliest Times to the Present as Told by the Explorers Themselves.* New York: Dial, 1952.

Swan, Robert. *Destination, Antarctica.* New York: Scholastic, 1988.

Time-Life Books. *Voyages of Discovery: Timeframe AD 1400-1500.* Alexandria, Va.: Time-Life Books, 1989.

Tinling, Marion. *Women into the Unknown: A Sourcebook on Women Explorers and Travelers.* New York: Greenwood Press, 1989.

Wallis, Helen. "'Things Hidden from Other Men': The Portuguese Voyages of Discovery." *History Today,* June 1986, 27-34.

Wilcox, Desmond. *Explorers.* London: British Broadcasting Corporation, 1975.

Wilford, John Noble. *The Mapmakers.* New York: Knopf, 1981.

ABOUT EXPLORERS AND SUBJECTS MENTIONED IN THIS BOOK

PART ONE

Alper, Ann F. *Forgotten Voyager: The Story of Amerigo Vespucci.* Minneapolis: Carolrhoda, 1991.

Bedini, Silvio. *The Columbus Encyclopedia.* New York: Simon and

Schuster, 1991.

Columbus, Christopher. *A Log of Columbus's Voyage to America in the Year 1492, as Copied Out in Brief by Bartholomew Las Casas.* Hamden, Conn.: Linnett, 1989.

Dodge, Stephen C. *Christopher Columbus and the First Voyages to the New World.* New York: Chelsea House, 1991.

Dolan, Sean. *Christopher Columbus: Intrepid Mariner.* New York: Fawcett Columbine, 1989.

Fernandez-Armesto, Felipe. *Columbus.* New York: Oxford University Press, 1991.

Goode, Stephen. "Debunking Columbus." *Insight,* Oct. 21, 1991, 10-17.

Gray, Paul. "The Trouble with Columbus." *Time,* Oct. 7, 1991, 52-59.

Jensen, Malcolm. *Leif Erikson, the Lucky.* New York: Franklin Watts, 1979.

Jungersen, Kenneth. "The Search for Vinland: On the Trail of the Vikings." *Oceans,* Sept.-Oct. 1985, 11-19.

Leon, George. *Explorers of the Americas Before Columbus.* New York: Franklin Watts, 1989.

Levinson, Nancy. *Christopher Columbus: Voyager to the Unknown.* London: Lodestar, 1990.

Lyon, Eugene. "The Search for Columbus." *National Geographic,* Jan. 1991, 2-39.

Meltzer, Milton. *Columbus and the World Around Him.* New York: Franklin Watts, 1990.

Nebenzahl, Kenneth, ed. *Atlas of Columbus and the Great Discoveries.* New York: Rand McNally, 1990.

Sale, Kirkpatrick. *The Conquest of Paradise: Christopher Columbus and the Columbian Legacy.* New York: Knopf, 1990.

Smith, Anthony, *Explorers of the Amazon.* New York: Viking, 1990.

Soule, Gardner. *Christopher Columbus: On the Green Sea of Darkness.* New York: Franklin Watts, 1988.

Stefoff, Rebecca. *Vasco da Gama and the Portuguese Explorers.* New York: Chelsea House, 1992.

Stott, Ken. *Columbus and the Age of Exploration.* New York: Franklin Watts, 1988.

Sudo, Phil. "Searching for the Real Columbus." *Scholastic Update,* Sept. 20, 1991, 2.

Taviani, Paolo Emilio. *Columbus: The Great Adventure.* New York: Crown, 1991.

Viola, Herman J., and Carolyn Margolis, ed. *Seeds of Change.*

Washington, D.C.: Smithsonian Institution Press, 1991.
When Worlds Collide: How Columbus's Voyages Transformed Both East and West. Newsweek special issue, Fall/Winter 1991.
Wilford, John Noble. *The Mysterious History of Columbus: An Exploration of the Man, the Myth, the Legacy.* New York: Knopf, 1991.

PART TWO
Berton, Pierre. *Arctic Grail: The Quest for the Northwest Passage, 1818-1909.* New York: Penguin, 1988.
———. "The Search for Franklin." *Reader's Digest,* March 1989, 171-200.
Blackwood, Alan. *Captain Cook.* New York: Franklin Watts, 1987.
Blassingame, Wyatt. *A World Explorer: Ponce de León.* Champaign, Ill.: Garrard, 1965.
Blumberg, Rhoda. *The Voyages of Captain Cook.* New York: Bradbury, 1990.
Brown, Warren. *The Search for the Northwest Passage.* New York: Chelsea House, 1991.
Cameron, Ian. *Lost Paradise: The Exploration of the Pacific.* Topsfield, Mass.: Salem House, 1987.
Forbath, Peter. *The River Congo.* New York: Harper & Row, 1977.
Hall-Quest, Olga. *Conquistadors and Pueblos: The Story of the American Southwest, 1540-1848.* New York: Dutton, 1969.
Haney, David. *Captain James Cook and the Explorers of the Pacific.* New York: Chelsea House, 1992.
Innes, Hammond. *The Conquistadors.* New York: Knopf, 1969.
Knoop, Faith. *A World Explorer: Francisco Coronado.* Champaign, Ill.: Garrard, 1967.
Naipaul, V.S. *The Loss of El Dorado.* New York: Penguin, 1969.
Scott, J.M. *Icebound: Journey to the Northwest Sea.* London: Gordon & Cremonesi, 1977.
Stefoff, Rebecca. *Marco Polo and the Medieval Travelers.* New York: Chelsea House, 1991.
———. *Vasco da Gama and the Portuguese Explorers.* New York: Chelsea House, 1992.
Struzik, Edward. "Quest for the Northwest Passage." *Canadian Geographic,* Aug.-Sept. 1991, 38-47.

PART THREE
Allen, John L. *Jedediah Smith and the Mountain Men of the American West.* New York: Chelsea House, 1991.

Bierman, John. *Dark Safari: The Life Behind the Legacy of Henry Morton Stanley.* New York: Knopf, 1990.
Cameron, Ian. *Mountains of the Gods: The Himalaya and the Mountains of Central Asia.* New York: Facts on File, 1984.
Forbath, Peter. *The River Congo.* New York: Harper & Row, 1977.
Goetzmann, William H. *Exploration and Empire: The Explorer and the Scientist in the Winning of the American West.* New York: Norton, 1978.
———. *New Lands, New Men: America and the Second Great Age of Discovery.* New York: Penguin, 1987.
Hopkirk, Peter. *Trespassers on the Roof of the World: The Secret Exploration of Tibet.* Los Angeles: Tarcher, 1982.
Huxley, Elspeth. *Livingstone and His African Journeys.* New York: Saturday Review Press, 1974.
Laycock, George. "Jed Smith the Pathfinder." *Boy's Life,* August 1987, 22.
Mirsky, Jeanne, ed. *The Great Chinese Travelers.* New York: Pantheon, 1964.
Ransford, Oliver. *David Livingstone: The Dark Interior.* New York: St. Martin's, 1978.
Stanley, Henry M. *Through the Dark Continent.* 2 vols. New York: Dover, 1988. Originally published 1878.
Stefoff, Rebecca. *Marco Polo and the Medieval Travelers.* New York: Chelsea House, 1991.
Ward, Geoffrey. "The Darkest Continent: When Henry Stanley Described Africa, He Described Himself." *American Heritage,* April 1991, 11-13.

PART FOUR

Cameron, Ian. *Mountains of the Gods: The Himalaya and the Mountains of Central Asia.* New York: Facts on File, 1984.
Fawcett, Percy H. *Exploration Fawcett.* London: Century, 1988. Originally published 1953.
Fleming, Peter. *Brazilian Adventure.* Los Angeles: Tarcher, 1976. Originally published 1933.
Hedin, Sven. *My Life as an Explorer.* Garden City, New York: Garden City, 1925.
Shankland, Peter. *Byron of the Wager.* New York: Coward, McCann & Geoghegan, 1975.

Index

Rebecca Stefoff has written more than fifty books for young adults, specializing in geography and biography. Her longtime interest in reading and collecting travel narratives is reflected in such titles as *Lewis and Clark, Magellan and the Discovery of the World Ocean, Marco Polo and the Medieval Travelers, Vasco da Gama and the Portuguese Explorers, The Viking Explorers,* and numerous books on China, Japan, Mongolia, the Middle East, and Latin America. Ms. Stefoff has served as editorial director of two Chelsea House series, *Places and Peoples of the World* and *Let's Discover Canada,* and as a geography consultant for the *Silver Burdett Countries* series. She earned her Ph.D. at the University of Pennsylvania and lives in Philadelphia.

Picture Credits

Antikvarisk-Topografiska Arkivet, Stockholm: p. 11; Bancroft Library, University of California, Berkeley: pp. 86, 98, 105; Bettmann Archive: p. 37; Bibliothèque Nationale, Paris: pp. 46, 66, 116, 125; Bildarchiv Preussischer Kulturbesitz, Berlin: p. 20; British Library: pp. 41 (C.46.i.18), 47 (photo from New York Public Library, Astor, Lenox and Tilden Foundations, General Research Division), 49 (Maps C.3.d.11), 50-51 (1C.9304), 54-55 (Maps C.2.d.7); British Library, Oriental and India Office Collections: pp. 94 (photo 392/8 0304i), 97 (OR 8210/P2); British Museum: p. 44 (photo from New York Public Library, Astor, Lenox and Tilden Foundations, General Research Division); Culver Pictures: p. 27; Florida State Archives, Tallahassee: p. 67; Fotomas Index: pp. 24, 32; Germanisches National Museum, Nürnberg: p. 56; Gilcrease Museum, Tulsa, Oklahoma: p. 100; Giraudon/Art Resource, New York: p. 30; Sven Hedin Foundation, Folkens Museum, Stockholm: pp. 134, 136, 137; Gwyn Jones, *A History of the Vikings,* Oxford University Press, 1968: p.13; Joslyn Art Museum, Omaha, Nebraska: p. 104; Library of Congress: cover, pp. 26 (LC USZ 62-59702), 33 (LC USZ 62-17888), 38 (LC USZ 62-33889), 40, 62, 63, 65, 69, 72 (LC USZ 62-25357), 77 (LC USZ 62-058158), 78 (LC USZ 62-22035), 83 (LC USZ 62-38672), 106, 114, 138; Library of Congress, Rare Book and Special Collections Division: pp. 103, 112; David Livingstone Center, Blantyre, Scotland: p. 110; The Mansell Collection: p. 111; Manuscript Institute of Iceland, Reykjavik: p. 12; Mary Evans Picture Library: p. 108; The Metropolitan Museum of Art, Gift of the Dillon Fund, 1973 (1973.120.3): p. 90; The Metropolitan Museum of Art, Thomas J. Watson Library: p. 68; Museo Naval, Madrid: pp. 34, 52-53; Museum für Gestaltung, Zurich: p. 93; National Archives: p. 71 (106-IN-2384CZ); National Maritime Museum, London: p. 76; National Portrait Gallery, London: p. 126; New York Public Library, Astor, Lenox and Tilden Foundations, General Research Division: pp. 22, 42, 58-59, 87, 88, 113, 123, 129; New York Public Library, Picture Collection: pp. 130, 132; New York Public Library, Astor, Lenox and Tilden Foundations, Rare Books and Manuscripts Division: p. 18; The Oakland Museum, Kahn Collection: p. 102; Pierpont Morgan Library: pp. 28 (M.525, f.17), 45 (M. 638, f.23v detail); Royal Society for Asian Affairs, London: p. 121; David St. Clair, *The Mighty Mighty Amazon,* Souvenir Press, London: p. 35; Scott Polar Research Institute, University of Cambridge: p. 84; Wayne Sturge, Department of Development, Government of Newfoundland and Labrador: p. 17; Tate Gallery, London/Art Resource, New York: p. 80; Universitetets Oldsaksamling, Oslo: pp. 9, 10; University of Maryland Baltimore County, Photography Collections: p. 124.

DISCARDED